THE BOMB DOCTOR

THE BOMB DOCTOR

A SCIENTIST'S STORY OF BOMBERS, BEAKERS, AND BLOODHOUNDS

KIRK YEAGER, PhD
& SELENE YEAGER

A REGALO PRESS BOOK

The Bomb Doctor:
A Scientist's Story of Bombers, Beakers, and Bloodhounds

ISBN: 979-8-88845-297-4
ISBN (eBook): 979-8-88845-298-1

Cover design by Conroy Accord
Interior design and composition by Greg Johnson, Textbook Perfect

As part of the mission of Regalo Press, a donation is being made to the EOD Warrior Foundation, as chosen by the authors. Find out more about this organization at: https://www.eodwarriorfoundation.org/.

Regalo Press
New York • Nashville
regalopress.com

Published in the United States of America
1 2 3 4 5 6 7 8 9 10

To my wife Deborah
who has provided stability in a life of chaos,
and my two sons Alec and Jared,
who, as they make their way into the world,
I wish the best in making their own marks.

Contents

INTRODUCTION

The Way of the Bomb

My sister and her family often visit over the holidays. A few Easters ago, they were greeted by a tall stack of sawed-off bamboo stalks drying on my front porch next to a bucket filled with roughly four thousand match sticks that had their heads clipped off and a metal toolbox labeled "Explosives Material" in Sharpie sitting off to the side. When she inquired about my little project, I assured her it was fine...just something I'd read about in *Inspire* magazine.

Seriously. It *was* something I'd read about in *Inspire*—an online magazine published by AQAP (al-Qaeda in the Arabian Peninsula). The magazine is an important brand-building tool for the organization. Like many magazines, it also features recipes—except these aren't for cakes and Instant Pot stews; these are for bombs. The edition I'd just finished before my sister came to call included an illustrated guide for harvesting explosive fillers for deadly pipe bombs from common match heads. I thought I'd give it some study. However, the first stage was harvesting explosives from thirty thousand matches.

After months of cutting off match heads, crushing match heads, grinding match heads, and wondering if my wife would notice if I borrowed her food processer to make a match puree, I was able to determine the sections of the recipe that worked and posed a viable threat. I was also able to see all the little things that did not work so well and would more than likely result in terrorist flambé. I won't be posting these on Pinterest anytime soon.[1]

Terrorists come up with their creative ideas for bombs by doing what many other inventors do. Rather than reinvent the wheel, they steal from history...and from each other, generally taking something used for construction and harnessing it for mass destruction. You can trace the origins back to black powder, which was used in cannons and powder kegs of the 1800s. Then, courtesy of Alfred Nobel, came dynamite (which produced an offshoot of putty-like explosives like gelignite).

Dynamite helped us build railways and mine for underground resources to make energy. But it also has the obvious potential for great destruction. Nobel himself would go on to make a fortune selling many of his explosive inventions for military application. But—in an odd twist of fate—before he died, a local newspaper printed his obituary by mistake. In it they referred to him as the "Merchant of Death." He was so shaken by this being his lasting public legacy that he created the now-famous Nobel Peace Prize. In a move that would have surely made Nobel proud, in 2017, the Nobel Prize committee granted that award to the International

[1] Important note: I'm not recommending—in fact I'm strenuously discouraging—that others read *Inspire*. Depending upon where you live, it can actually land you in a fair amount of hot water. Britain and Australia have strict anti-terrorism laws that can make it a crime to download its contents. Dozens of people have been arrested and prosecuted in Britain just for downloading *Inspire*.

Campaign to Abolish Nuclear Weapons, which are, of course, the most devastating bombs known to man.

It used to be harder for would-be bombers to find new ways to wreak their havoc: word of mouth only spreads so far, and books aren't exactly current resources. Today, thanks to outlets like *Inspire*, novel ways to conjure up explosive devices are just a click away. In the 2000s, we saw a wave of pressure-cooker bombs—kitchen appliances packed with shrapnel and detonated by electronic devices—including those planted by convicted bomber Ahmad Rahami in New York City's Chelsea district in the fall of 2016, said by friends to be informed by the likes of *Inspire*. There is no shortage of angry minds in search of diabolical ways to harness any and every volatile reaction.

The FBI Explosives Unit fields countless calls from commercial airlines, government officials, and concerned citizens asking if everyday, seemingly innocuous household goods like toothpaste or mouthwash could be used to explode a car, airplane, or building.

People have made explosives out of plenty of chemicals that are used in common consumer goods, including fertilizer, Hydrogen Peroxide, and paint thinner. Richard Reid, the would-be British shoe bomber, tried unsuccessfully to discharge a plastic explosive he had stuffed into the waffle pattern in the sole of his shoe using a detonator made with one such chemical (Triacetone Triperoxide or TATP) during a Paris-to-Miami flight in 2001. That was the first major terrorist attack against a commercial flight destined for the United States since the Pan Am Bombing in 1988 and is the reason we remove our shoes in airports to this day.

Several years later, counterterrorist authorities reported thwarting as many as ten potential terrorist attacks on London-to-US flights where would-be bombers planned to mix a volatile solution using one of the chemicals needed to make TATP. This is why you can no longer bring large quantities of liquids through airline security.

The forensics my colleagues and I do after the bomb drops is designed to clear the way for preventing future tragedy. When you understand the bomber's materials and motivations, you can foil them before they light their first fuse. The challenge is staying one step ahead of them, being able to see far enough ahead to block their move. It's like a game of chess, only instead of having Rooks, Queens, Bishops, and Knights moving predictably side to side and along diagonal lines, someone introduces a Flying Rhino that can wipe out a whole row. You need to see that coming.

To stay one step ahead, I read what the bad guys read and have spent a good deal of my career manufacturing explosives that replicate the materials deployed by terrorists. My job is to understand what these folks are cooking up and to help the good guys safely make it go away.

It is not always without folly. I need to keep my own ego in check, lest the materials I am working on do the job for me. Early on in my FBI career, I made a training film for bomb techs with an engineer colleague of mine. This gentleman later went on to lead the entire FBI unit that does our forensic bombing investigations. In the film, we were producing a generic explosive device often made by juvenile bombers. It starts out with two chemicals placed in a plastic two-liter soda bottle. You add water to start a chemical reaction, which releases gas and slowly pressurizes the bottle. Eventually enough gas is produced to pop the bottle and, in its most common application, blow up a household mailbox.

Knowing how finicky these devices can be, my colleague and I were very cautious in the amount of chemicals we added. Using just the minimal amounts of the materials, we closed the bottle cap and waited. And waited. Eventually it became obvious to us that we were too timid in our construction. At this point, we should have incrementally increased the amount of material being applied to the task at hand. However, I think we both took the lack of reaction as a professional affront.

Neither of us was going to let the top FBI bomb specialists fail at making a bomb a twelve-year-old delinquent could produce.

In an effort to keep from providing bomb-making instructions to the reader, I will not mention the chemicals used. However, one of them resembles tinsel. In our first device, the bottle had only a small mound of tinsel sitting on its very bottom. By the time we had the second device put together, it looked like someone had crammed a whole Christmas tree into the bottle. I knelt down with a funnel and added water. And then...*HISSSSSSSS*—a very immediate "oh shit" moment.

For a brief second, I caught my partner's eyes as we simultaneously surmised it would be prudent to de-ass the area. He was standing and able to move faster.

Being on one knee, it took me a second longer to backpedal. In this crucial second, the mixture was generating gasses so rapidly that the bottle was shaking and frothing. As I stood, it toppled over, its mouth facing me. We never had a chance to put the lid on the bottle top, which left the perfect opening for all the hot water and nasty chemicals to hose me down as the bottle became a projectile, rocketing over a hundred feet into the air.

Thankfully, I was wearing safety gear, and we poured water over me to wash away the caustic chemicals, so I did not get too badly injured. But I was reminded of one of the adages passed onto me by the FBI agent who first trained me in the lab: "If you are going to run with wolves, make sure you don't trip and fall." I have been lucky enough to only stumble a time or two.

Fortunately (though not for them), the bad guys often have two left feet.

Equipped with just enough knowledge to make them dangerous to themselves, their stumbles frequently result in receiving a hard lesson in shock physics before the bomb makes it to its intended destination. Other times, as in the case of Cesar Sayoc, the bomb thankfully doesn't go off at all.

When a caretaker at billionaire George Soros's house went out to get the mail on Monday, October 22, 2018, one of the packages didn't look quite right. It was a taped-up eight-by-ten envelope that had a bit of heft. Not a typical piece of parcel post. The caretaker smartly placed the mystery package in a stand of trees away from the house before summoning law enforcement.

Upon tearing it open, the authorities reportedly found an unwelcome gift nestled inside the bubble wrap interior: a six-inch pipe along with a battery, wiring, a small clock, and a black powder.

A bomb.

The original weapon of choice among the "Davids" of the world who want to bring down their perceived Goliaths. It's a theme that has run through bombing campaigns since humans learned to harness the powers of explosive chemical reactions. Lone vigilantes or cells of like-minded angry individuals use high-powered explosives to amplify their stature and take down those they feel are their larger, oppressive enemies.

Zeroing in on Cesar Sayoc—the fifty-six-year-old former male dancer and pizza deliveryman from Aventura, Florida—proved to be a fairly straightforward forensic affair. For one, he didn't exactly hide his raging animosity as he drove around town in a white stretch van that he'd plastered with giant headshots of public figures like Michael Moore and Hillary Clinton placed in crosshairs. More importantly, his devices, which he sent to the homes of high-profile figures including Barack Obama, Joe Biden, and the Clintons, never blew up in the hands of their recipients.

Instead, the FBI was able to recover more than a dozen devices—all intact.

That's a veritable gold mine of forensic information. It took very little digging to discern how the devices were constructed, what materials were used, and how they were packaged and postmarked. He essentially left a trail of breadcrumbs leading the Feds to his doorstep.

According to news reports, Sayoc consistently misspelled the names of some of his targets. He misspelled Representative Debbie Wasserman Schultz's last name as "Shultz" on both the package containing her pipe bomb and in Twitter rants from an account believed to be his. Similarly, he'd misspelled Hillary as "Hilary" in his social media diatribes and on the pipe bomb sent to her. A DNA sample found inside a package sent to Obama matched Sayoc. He'd also left behind a fingerprint on the envelope mailed to Democratic representative Maxine Waters of California that matched samples taken from Sayoc during a previous arrest. It was a relatively easy case. Classic CSI, really.[2]

In reality, it's rarely ever classic CSI.

Why? Because bombs often blow up. And when they do, we're not left with neat packaging riddled with easy-to-inspect evidence. The forensic work required to solve the mystery of a crime involving high explosives is like nothing you've ever seen on a made-for-television investigative series.

Forget about fingerprint evidence—that's blown to smithereens. There is rarely DNA of any value. As a true bomb detective, what you have to work with are fragments, soot, fields of twisted metal, and charred human remains. You have carnage and chaos. And in that sea of wailing sirens, beeping horns, screaming survivors, and the stench of diesel fuel and decaying bodies, your job is

[2] Sayoc eventually pleaded guilty to mailing sixteen improvised explosive devices to victims across the country. He was sentenced to twenty years in prison. According to news reports he blamed his behavior on mental illness and excessive use of steroids. He also claimed that though the devices looked like bombs, he did not intend for them to explode.

to ferret out forensic clues in a type of macabre scavenger hunt. Your mission is to find what you need to reconstruct the scene, recreate the explosive device (or devices), and determine what the bomb looked like and what went down before it was all torn asunder, in the hopes of ultimately bringing the bad guys to justice and preventing further attacks.

That's what true bomb forensics is like. You're walking into hell—blindfolded. You don't know what's in front of you; you don't know where the path will lead you. You just start pursuing different avenues, wading through idle speculation, and finding forensic clues to slowly develop a fuller picture.

This process does not happen overnight. It takes weeks, months, years even. But it's work we learn from, and work that helps prevent more catastrophes in the world. It's gritty. It's gruesome. It's time consuming and sometimes dangerous. And it's 100 percent worth it.

CHAPTER 1

Mass Destruction for the Masses

During my tenure as a research science and adjunct professor of chemistry at the New Mexico Institute of Mining and Technology—before I officially joined the Bureau—I conducted field tests in Socorro for the FBI. We were doing a deep dive into the use of Ammonium Nitrate (AN)—a type of fertilizer mixed with a variety of fuels to make car bombs, which were wreaking havoc in the US and UK at the time. Governments on both sides of the pond were eager to learn more about these materials and find ways of preventing terrorists from using them to create mass mayhem. Over the course of six years, my team and I prepared approximately 128,000 pounds of explosives created from fertilizer and a wide assortment of fuels. I cannot recall the number of old junker vehicles I disseminated into the desert during that time frame, but one test shot in particular stands out.

During one of our largest tests, meant to simulate a truck bomb, we piled about four thousand pounds of AN—which had been mixed with diesel fuel to produce an explosive called

ANFO—on a testing pad. Everyone else watched the pad through a periscope-like assemblage of mirrors from the blast-proof observation area deep within the bunker. The doorway faced away from the range, so fragmentation was not of concern. It was spring and I was enjoying the cool mountain desert air. I was wearing a ball cap and the technician next to me had on a wide brimmed straw cowboy hat. I remember hearing the countdown.

"Three, two, one..."

A peculiar moment of stillness accompanied a huge fireball and blast, as if captured on a reel of a silent film. As the shockwave eventually reached out and touched me, as if two huge hands had thumped me from the front and back simultaneously, my ball cap snapped from the brim facing forward to standing straight up at attention, much like Daffy Duck's beak after being shot in the face. My colleague's cowboy hat blew off his head and sailed twenty feet into the far corner of the bunker.

But witnessing the forces of a large-scale explosion in a controlled field test is far different than the horror the world had recently witnessed on the streets of Oklahoma City, when massive amounts of this type of explosive found themselves in the wrong hands.

It had been an otherwise normal, clear sunny day in Oklahoma City when Timothy McVeigh drove a truck laden with approximately two tons of fertilizer-based explosives and detonated it outside the Alfred P. Murrah Federal Building. Within a fraction of a second, a wall of superheated pressurized gases smashed into the building and shattered its facade. This wall of pressure pushed upwards against the floors, lifting them against gravity (a direction architects and engineers never planned for or designed against). As the blast wave swept by, the floors relaxed, only now the structural supports holding the floors up against the pull of gravity were no longer there. They'd been pulled apart by the upward force of the blast and could no longer bear the weight when the floors settled back down. As the floors collapsed, they created a massive gaping hole in the building, which is the image we're so familiar with today.

Ultimately, the blast leveled one-third of the nine-story concrete building, as well as damaged 324 buildings within a sixteen-block radius; it killed 168 people, including 15 children in Murrah's day care center.

McVeigh, a Gulf War veteran enraged over the handling of the Waco siege, wanted to inspire a riot against what he saw as a tyrannical federal government; he defended the bombing as a legitimate act. That is what bombers do. It's what they have always done, though the present has a way of forgetting the past.

We tend to think of terrible acts as being unique to the here and now, a sign of deteriorating times. It's part of human nature to romanticize a simpler, better past. But Timothy McVeigh was not the first US bomber to make his seething rage against the government known through the most macabre of means. He was also not the first to include children in his body count.

In 1927, Andrew Kehoe conducted a multi-prong attack against a schoolhouse and its administrators in Bath, Michigan. Over a period of months in the spring of '27 Kehoe used his position as a trusted handyman to pack crawlspaces in the newly constructed Bath Consolidated School House with hundreds of pounds of dynamite and the explosive Pyrotol. The night before the attack, Kehoe killed his wife and rigged all the buildings on his farmstead with incendiary devices. At 8:45 on the morning of May 18, the last day of school for that year, timed detonators initiated the charges underneath the north wing of the three-story schoolhouse. The resulting explosion collapsed the structure and killed thirty-eight elementary school children and six adults.

Only part of the attack worked to plan. Half of Kehoe's charge failed to initiate; the police later recovered over five hundred pounds of explosives with the assistance of a fourteen-year-old boy small enough to fit into the crawl spaces where they were secreted.

In a final act of savagery, Kehoe packed his car with all the heavy metal tools from his barn and a few remaining cases of dynamite. He set off the incendiary devices in all his buildings and drove off to the smoldering remains of the school.

Fire trucks responding to the massive fire on his property passed Kehoe as he drove to his final destination. When he got to the school, Kehoe called the superintendent, who was helping pull bodies from the wreckage, over to his car. He reached over and pulled out a rifle, with which he shot into the cases of dynamite.

The explosion killed him and the superintendent. Thus, he became the first suicide car bomber in the US, as well as the first to target children.

Kehoe's attack was the most deadly bombing in US history at the time. It would remain so until Oklahoma City. Reasonable people try to understand the root of such evil. They ask "Why?" Like McVeigh, Kehoe felt wronged by the government, in this case for increasing taxes on his farm and denying his bid for township clerk. When disturbed people like Kehoe and McVeigh become incensed over a real or imagined wrong, and perceive the perpetrator of this injustice is too strong to take on, they often turn to high explosives as the great equalizer. That is the nature of bombers and their campaigns.

At the time of the Oklahoma bombing, I was at a terrorism symposium on the campus of New Mexico Tech. A young post-doctoral researcher learning about explosives science and range testing, I had been at "Tech" for about sixteen months. During that time, I had embarked on a career of replicating terrorist explosive recipes and conducting detailed studies on the products they created. Tech was unique in that it contained a research division dedicated to the study of explosives (the Energetic Materials Research and Testing Center, EMRTC). It also contained a cadre of some of the best-known explosive researchers, scientists, and engineers whose playground covered forty-two square miles of the New Mexican desert.

On the morning of April 19, 1995, I was sitting in Tech's conference center attending EMRTC's explosive industry research conference. The theme of this particular conference was "Terrorist Application of Explosives." Through contacts from past research programs with the FBI, the FAA, and the UK government, we had a star lineup of speakers, all of whom dealt with terrorist attacks at the national level.

First up at 8:00 AM was the FBI briefing on the 1993 World Trade Center bombing. Just six months earlier, I had met the FBI team who investigated the bombing. The team had synthesized a massive amount of Urea Nitrate years earlier but were forbidden to test it by the courts while the trial was going on. Once the guilty verdicts came in, the FBI found itself with 1,200 pounds of improvised explosive they really didn't want sitting around and a strong desire to blow it up.

They came out to EMRTC to do just that.

Fast forward six months to the conference. The agent in charge of the bombing forensics captured the audience's attention with the investigation backstory. A terrorist cell produced half a ton of explosives in an apartment building, assembled a devastating vehicle bomb (complete with dynamite made from toilet paper rolls, Ammonium Nitrate, and Nitroglycerin), drove it into the B2 level of the World Trade Center parking garage, lit a fuse with a twenty-minute delay, and then barely got out in time in a getaway car. The agent explained how the bomb created a crater sixty yards across in the interior of the building that punched holes above and below through multiple stories. Out of the massive heap of cars tumbled into the underground crater, investigators pulled the magic piece of the bomb vehicle containing the critical vehicle identification number (VIN number). The VIN number discovery led to a Ryder rental agency in New Jersey. Agents were informed that the individual who rented the bomb vehicle had attempted multiple times to get his four-hundred-dollar deposit back (claiming the truck had been stolen). The third time he came back he was provided with more than he bargained for: FBI agents manning the counter.

By all accounts, the EMRTC conference was starting off with great potential. But at 9:02 AM, one time zone and approximately six hundred miles away, a new horror was unraveling, engraving itself into the annals of American history.

Shortly before 9:00 AM Oklahoma time on April 19, 1995, just as the FBI speaker was stepping on the stage, Timothy McVeigh was sitting at a stop light in Oklahoma City. As he waited for the light to turn green, he lit a time fuse with a two-minute delay. This fuse ran into the bed of his Ryder rental van, which was packed with multiple fifty-five-gallon drums containing two to three tons of fertilizer-based explosives. When the light turned green, he drove forward and parked the van in front of the Alfred P. Murrah Federal Building, got out, and walked away from the scene to his getaway vehicle. And then the van exploded.

Some words have their meaning so deeply ingrained in our being that it seems pointless to define them. Even though the concept of an explosion seems so primal and fundamental, few realize the chemistry and physics that combine in a vortex of chaos to create this event.

As a scientist I would explain an explosion as "a rapid expansion of matter into a greater volume." As an author, I defer to explosive expert Tenney L. Davis's words from his 1940s textbook: "It seems unnecessary to define an explosion, for everyone knows what it is—a loud noise and the sudden going away of things from the place where they have been." None have said it better.

Explosions happen in less than the blink of an eye, but I'm going to slow them down for you here. In short, explosions are pure, simple energy transfers. The source of this energy is a rapid combustion reaction. A typical explosive turns from a solid or liquid into superheated gases in a thousandth of a second. As a rough estimate, every gram of explosive produces a liter of gas upon detonation. This means a shot glass of explosive produces over seven gallons of gases. Timothy McVeigh's bomb unleashed in the neighborhood of twenty million gallons of gas (almost enough to fill twenty Olympic-size pools).

Confined gases develop pressure when they are heated (which is why throwing an aerosol can onto a fire is not a highly recommended recreational activity). In the case of Oklahoma City, the tons of fertilizer-based explosive were transformed from many fifty-five-gallon drums of solid particles into gaseous products at thousands of degrees in temperature. These gases produced about ninety-five thousand atmospheres of pressure. These gases expanded violently against their environment and tore to pieces everything in close proximity to them. The drums were ripped to small shreds of plastic, some of which were found on rooftops a hundred yards away. The truck carrying the bomb was torn to pieces, with the rear axle containing the critical VIN stamp found two blocks from the seat of the blast. Physical items torn and propelled by an explosion are referred to as fragmentation (or "frag" in lay terms).

It is difficult to think of the air that surrounds us as a potential devastating force. However, hurricanes and tornadoes show all too clearly the power moving air can exhibit when given the proper impetus. Tornadoes pale in comparison to the power of an explosion. When the expanding gases created by the detonation of the Oklahoma City bomb hit the surrounding atmosphere, they compressed the air and sent a condensed plug outwards like an invisible tsunami. This pulse of pressurized air, referred to as the blast wave, creates an effect referred to as air blast.

The closer to the seat of the blast, the more powerful the blast wave. Where the actual explosive makes contact with its container, the pressure produced at the explosive surface can reach thousands of pounds per square inch (psi). By way of comparison, the atmospheric pressure of the earth at sea level is about 14.7 psi. The pressure pulse can do tremendously destructive work. At 300 psi, holes are punched in the ground the bomb sits next to, producing craters. At 100 psi, enough damage can be done to the human lung to induce death. At 1 psi, most people will be knocked off their feet. To provide perspective, at 1 psi the typical four-foot-by-six-foot dining room table experiences about 3,500 pounds of force. An average male (with total body surface area of 3,000 square inches) experiences roughly 1,400 pounds of force.

The Oklahoma City bomb generated pressures of 1 psi at distances of over two football fields away from the blast site. The sections of the Murrah building within fifteen feet of the bomb saw pressures in the range of 9,600 psi. For typical construction, most buildings will be totally destroyed if exposed to pressures of 10–12 psi.

The lethality of an explosion comes from the trifecta of explosive effects: air blast, fragmentation, and thermal pulse. Air blast is devastating close-in but dies off quickly with distance, and the thermal pulse of an explosive is short-lived, and is only felt in close proximity to the blast. But the true lethality of a bomb lies in the fragmentation. The accelerated physical items propelled from close proximity to the bomb possess the greatest potential to maim and kill at the furthest distances. Looking at a bomb is akin to looking down the barrel of a shotgun.

The blast, thermal effects, and massive array of fragmentation created by the truck bomb in Oklahoma City would wreak devastation for blocks.

Back at the EMRTC conference on that fateful day, with an audience full of top bombing experts, pagers started to go off only minutes after the blast, and it became obvious something huge had transpired. Within the span of two hours the experts streamed out of the building, determined, hurrying to get to work. Little did I know that, only five years later, I, too, would be called upon to help bring to justice those responsible for similar atrocities.

Over that five-year span, I expanded my testing of unconventional explosives. I became the "go to" scientist for numerous government agencies, often called upon to make and study the broad range of concoctions being brewed by terrorists across the world, including those using Ammonium Nitrate. As years passed, I made more and more dangerous materials for a wide range of organizations. And in doing so, I gained a reputation as what some referred to as an "explosive guru."

I wish I could say that we eliminated Ammonium Nitrate as a potential threat. We haven't. At the writing of this book, a disturbed Pennsylvania man blew up himself, his son, and a friend in a homemade car bomb made partially with this fertilizer. Ammonium Nitrate remains a major player in the battlefields of Afghanistan. But thanks to the public awareness following Oklahoma City, buying massive building-leveling quantities is harder to do without being questioned by sellers. And you'd raise eyebrows (and the interest of the authorities) if you tried to purchase it by the ton.

CHAPTER 2

Make a Bomb in the Kitchen of Your Mom

One of the great appeals (and obvious perils) of explosives to those with malicious intent is that you can make them yourself. Indeed, one of *Inspire*'s most popular features is "Make a Bomb in the Kitchen of Your Mom."

Catchy title, but al Qaeda followers were hardly the first terrorists to this precarious potluck. Counterculture icon William Powell brought his bomb recipes (along with a few for mind-altering hallucinogens) to the party in 1971 with *The Anarchist Cookbook*, which, as promised, contained any number of recipes for disaster.

I was just six years old when that first cookbook rolled off the press, but I began analyzing so-called kitchen sink bombs well before they became an Internet sensation among the deranged. During my time as a young research scientist at New Mexico Tech's Energetic Materials Research and Testing Center, the

Federal Aviation Administration (FAA) wanted to learn more about the threat posed by the exotic explosives contained in the pages of the Anarchist Literature. Many had speculated about their capabilities, but very little science existed to shed true insight.

It didn't take me very long to demonstrate that common, easy-to-access household chemicals that line our cupboards and closets—stuff like Hydrogen Peroxide and paint thinner—could be used to produce explosives like Triacetone Triperoxide (TATP), and with devastating effects. Around this time, Palestinian militants began experimenting with the use of TATP as well. They found this explosive material so devastating and unstable (often killing the would-be killer before he could do ill deeds) that they named it "Mother of Satan." Thankfully my experiences with TATP were not as dramatic as those of the less chemically skilled terrorists.

Fortunately (relatively speaking, anyway), most of the world's scratch bomb bakers often blow themselves up before they mess up too many other people's lives. That's because they have only just enough knowledge to be dangerous. Take Benjamin Morrow, for example. The twenty-eight-year-old from Beaver Dam, Wisconsin, thought it would be fine to have thirteen medium-size jars of various states of TATP hanging around the kitchen while attempting to cook up an unknown explosive mixture on his stove. Regardless of his intentions he succeeded in making a bomb...and on March 5, 2018, he accidentally detonated it and blew himself up at the stove. The place was so packed with unstable materials intermingled with copious debris from the collapsed ceiling that the authorities couldn't safely clear what remained and had to simply burn the place down to the ground ten days later. The moral here is that overconfident people who play with unstable substances sometimes blow themselves—or at least parts of themselves—up.

Once, while enjoying a rare quiet moment in my office, which I knew by experience would not last too long, an FBI bomb tech called to inform me that he was at the residence of a gentleman who had showed up at the hospital emergency room sans three of the digits on his right hand. This physical state piqued the doctor's interest and resulted in a series of questions about how the patient came to be without said body parts.

Explosions are unique in what they do to a body. Losing fingers to a bladed device, like a lawn mower, for example, does not resemble having them literally torn off a hand by blast pressure. Under pressure, just slightly less than the blast he survived, the man admitted that he had been experimenting with explosives in his mom's kitchen and had experienced an unintentional energetic event. (Unsurprisingly, this was not the first explosive accident for this man. Years earlier, he had lost segments of fingers in another, less severe, explosive experiment, and he'd been able to convince the doctor at the time that it was the result of some benign household appliance, like a Vitamix smoothie gone terribly awry.)

Unfortunately for the man, his fingers, left back at the scene of the accident, could not be reattached, as his brothers—assuming their mother would be rip-shit angry about the mess in the kitchen—had dutifully cleaned up the blood splatter and disseminated flesh. They flushed the fingers down the toilet along with any hope of restorative surgery.

Even when people successfully cook up their concoction, the train nearly invariably goes off the rails in ways I could never see coming.

Such was the case with another budding bomber (let's call him Homer) who successfully produced a highly volatile explosive called Hexamethylene Triperoxide Diamine, or, simply, HMTD. A chemical cousin to TATP, HMTD shares the same attributes, and making it can be extremely dangerous, as it can go off with little to no warning.

Homer didn't just successfully make the stuff; he had a stash in his house just waiting for him to come up with a creative way of utilizing it. And where did he stash it? Where seemingly every explosive maker keeps their goods: the beer fridge.

On a fifth day of May, Homer decided to experiment with another reactive mixture he had just lying around—thermite. I remember the day because the traditional celebrations and libations that surround Cinco de Mayo foreshadow the chain of events that followed, which resulted in the local bomb squad giving me a call requesting advice on how to clean up the mess they discovered. Had Homer just stayed with thermite, it would have been an uneventful day. Thermite, comprised of Aluminum powder and Iron Oxide, does not explode. It burns white hot, spewing forth a volcanic cascade of metal and creating molten iron, which melts through anything gravity brings it in contact with. Ironically, thermite is a bitch to get to burn. But once it starts burning, you cannot put it out. Throwing water on it results in superheated steam and white-hot iron flying back at you. Putting sand on it does nothing, as the oxygen in the mixture keeps fueling the fire.

Homer, trying various ways to light the thermite, was left only with a pile of rust-colored powder, with residual material he had already lit on fire smoldering on top of it. While the heat from the burning material was not enough to set off the thermite, it was still quite warm.

Eventually, Homer remembered the HMTD sitting around in the beer fridge waiting for a chance to shine. Its moment had come. A shot glass filled with HMTD has enough energy to punch a hole in a dining room table upon explosion. Homer had enough to fill a portion of a mason jar.

HMTD requires almost no heat to get it to detonate. Homer decided that, if he dumped some HTMD on top of the thermite and lit the HMTD, it would surely get the material started. Homer poured the HMTD by hand on the thermite, which was still smoldering from the failed initiation attempts. Upon hitting the hot chemicals, the HMTD detonated, blowing off his hand.

Typically, I get called after someone blows off a hand in the attempt to make the material. Never had I seen someone try to use it to start a fire. It would be like using a stick of dynamite to light a cigar.

In cases of premature explosive initiation, we come in and clean up the mess after the budding chemists have done their destruction. But sometimes we have to come in and clean up their messes before they blow up others as well.

A few years back I received a call from Cathedral City, California, about a young man who, heavily under the influence of drugs, had become convinced that something was after him. He wasn't sure what—something not human, definitely something evil.... We'll never really know, nor I imagine does it really matter. Bomb techs entering his mobile home found he had made some flash powder—a very sensitive material—and dispersed it in a two-inch layer throughout the place. He'd dumped it all over the kitchen floor and down the shag carpet the entire length of a narrow hallway, into his bedroom, and all over his bedspread.

The plan was that, when this evil thing came for him, the man would Wile E. Coyote the powder with a match and then, like a cartoon, jump out the window. Except, I assure you that if you drop a flame on flash powder, you're not jumping out the window. You're blown out the window.

The police had managed to apprehend the individual in question and charged him for suspicion of possessing a destructive device—a pipe bomb—and materials intended for manufacturing a destructive device. He ultimately pleaded guilty on nine felony counts; he couldn't cause any more trouble, at least for the foreseeable future. But the question remained: what to do with the ticking time bomb of a trailer? After an initial inspection, public safety officials deemed the place too dangerous to enter again. Towing the trailer away would be akin to hauling a bomb down the highway.

I was called in to, metaphorically, defuse the situation. My mind went through the options. Gels and foams, like the kind firefighters use, all react with this stuff; using them is unpredictable and dangerously unstable. Water is often a great solution (though not always).[3] Running fire hoses into all the windows and flooding the place would eliminate the immediate risk. But draining those kinds of toxins into the lawn is largely frowned upon (and quite illegal) in the Golden State. That left one choice: carefully surround the whole trailer with non-flammable barrier walls and professionally burn it to the ground.

No doubt, learning that your neighbor may be an off-his-rocker Betty Crocker intent on producing mayhem is unsettling. As it should be. It's also unsettling how easily these bad actors can obtain ludicrous quantities of suspicious materials without raising red flags or having someone block their way or at least alert some authorities.

Case in point: the British terrorists who made backpack bombs that they exploded in three trains in the London Underground, as well as on a double-decker bus, killing fifty-two people and injuring seven hundred, bought out all the Hydrogen Peroxide in a local beauty supply house multiple times without anyone asking why. Najibullah Zazi, who plotted to blow up the NYC subway system, bought a dozen bottles of hair bleach. When the drug store clerk asked what he was going to do with it all, he just

[3] Fun fact about water and some explosives: some chemical combinations actually will sit happily until touched by a drop of H_2O, at which time they will react as if someone dropped a match in a drum of black powder.

said he had lots of girlfriends. Neighbors of the Brussels bombers, who coordinated three suicide bombings in Belgium, noticed a strong chemical odor coming out of their apartment as they cooked over *a hundred pounds* of TATP. No one ever made a peep.

But there's an easy way for concerned parties to alert the proper authorities safely and swiftly. The FBI has robust outreach and training programs designed to educate stores that sell chemicals bomb makers covet. I helped design many of them. And the FBI has a special hotline people can call if they encounter any suspicious purchases or activities. Like the New York City subways remind you: if you see (or smell) something, say something.

But outreach is a constant challenge in this dynamic environment. There are thousands of workers at hardware stores, big box outlets, and beauty supply stores across our country. They constantly come and go.

CHAPTER 3

Rage Against the Infernal Machine

The car in question was out of place. That's the one thing that stood out among the expansive wreckage of the biggest bomb blast in an urban area I had ever seen. It was one of my first major vehicle bomb cases. Or, I should say, bombings, plural. The car of note was part of the catastrophic damage left behind by an unprecedented bombing campaign in the Indonesian archipelago of Bali. This campaign included: a car bomb explosion that destroyed a crowded nightclub, killing 202 people; a second bomb inside a bar nearby; and an improvised explosive device (IED) detonation in the vicinity of the United States Consulate.

In my job, I have two missions that require a collection of all the evidence I can find. With the gathered evidence, chemists can determine what explosive was deployed in the bomb and, second, determine what the bomb was comprised of and how it functioned. For many other crimes, piecing together the evidence is fairly rudimentary. Dead body, check. Shell casing, check. Conclusion: I believe this man has been shot. Well, maybe it's not that simple, but my point remains. In an average bombing, it can take a long time to put a coherent picture together. Things are often not as they initially appear.

Oftentimes, it takes days to locate critical evidence. The larger the bomb, the further things fly, and the more time it takes. In Bali, we didn't have just a basic bombing. We had a massive vehicle bomb staged down the street from a suicide bomb, and a third bomb that went off miles away. Each one required meticulous attention to detail. This would have been complicated enough, but upon arrival I was told by some colleagues that fellow bomb experts brought in from Australia had discovered evidence of what they believed was the scene of a not-yet-identified fourth bomb attack. Outwardly I just nodded my head and looked over at my partner, Mark, the forensic lead on this outing and a seasoned bomb tech and forensic examiner.

We travel in teams, and Mark Whitworth and I were paired up on this journey. Mark knew me from my New Mexico Tech days and was one of the best forensic examiners I ever worked with. A true Southern gentleman with enough common sense to carry any day, I was fortunate to be teamed with him. I later became the Chief Explosives Scientist; Mark would go on to be the chief of the entire Explosives Unit. If he ever writes a book, buy two of them.

The journey to this crime scene could fill a chapter all its own, but suffice it to say that so far we had been deployed with no contacts in country or plan of attack for when we arrived, traveled thirty-three hours, arrived without visas, successfully avoided having to bribe our way into country, and been abandoned to our own devices, which, for Mark, included shoving a wad of cotton from an aspirin bottle up his nose to stem the flow of an ill-timed travel-induced nosebleed that hit just as he was mistakenly pushed back by the local police guarding the crime scene. I don't know what was running through his mind as we heard about bomb number four, but all I could think was "This shit is starting to lose its comic appeal."

We finally met our Aussie counterparts as we prepared to go view the mystery car together. I was introduced first. The response from one of the younger Aussie bomb techs was, "Doctor Yeager? *The* Doctor Yeager? I've heard all about you." I smiled back and shook hands, thinking, "Here we go again."

Frankly, the reception I often receive gets embarrassing. I publish a great deal for bomb techs, who have been brought up in their professional careers being taught from my works. They always look shocked when they meet me, giving one of two standard responses.

Response 1: "You're not what I expected." I am never really sure how to take this one.

Response 2: "You're a legend." Baffles me even more. My standard retort is, "Don't worry, I am much less impressive in person." Enough of my colleagues have been around when this exclamation is issued that it is a source of constant joy for them to mock it. One variant that I heard, however, was dead on. Years ago, a young bomb tech was introduced to me and he got "that look" in his eyes and blurted out, "Dude, you're an urban legend." I can own that.

Due to this distorted view others hold of me, any wild-ass speculation that I make will be taken as gospel. With my own team I can bounce theories about a bombing back and forth with casual disregard. I cannot do that with people who take everything I say as profound.

Back to our introductions to our Aussie counterparts. I acknowledged that I was "The Doctor Yeager." Mark was left to introduce himself without the fanfare.[4]

---

[4] The following day when an AFP chemist greeted me with the same "The Doctor Yeager!", Mark introduced himself with the added caveat, "You've never heard of me." Smart ass.

At the scene of the purported fourth bomb, the younger of the Aussie bomb techs was seemingly a little awestruck to be in our presence; the older one seemed more skeptical. In some ways, when old bomb experts meet there is a little circling and testing each other out, like fighters in a ring. Staring at us like a schoolteacher and motioning to the decimated bar down the road and the devastated nightclub, the older chap asked, “So what do you think happened here overall?” This was a test. I put on my pensive face, scowled a little for effect, and, with sweeping arms, I provided the dramatic summary.

“What I think occurred here is that a suicide bomber went into the bar down the block to detonate a device with two purposes. First, he wanted to kill as many patrons as possible. Second, he wanted to drive the survivors out into the street so they could be herded into the killing zone of the second, larger vehicle bomb.”

My hunch was rewarded when the older bomb tech smiled slightly and said, “Good. You passed the first test. You can think like a sick fuck. Now we have something we want to show the two of you.”

A couple hundred feet from the site of the vehicle-bomb attack, they proudly showed us the car they had “discovered.” It had been smashed down and mangled to hell in the middle. That in and of itself wasn’t so odd. What *was* odd, though, was that, while the car was parked along the street with cars to the front and back of it, it was far more pummeled and wrecked than any of the cars around it, which appeared relatively normal. Explosives can do strange things, but they can’t violate the laws of chemistry and physics. As the distance from a bomb increases, the damage it does to things decreases. This car was an enigma.

The Australian bomb techs sent to help investigate the scene were convinced a bomb we had yet to find had damaged the car. I wasn’t so sure. As I was relatively green on the ground with these types of investigations, I had to convince two seasoned techs that a scientist might have a better analysis of the scene.[5]

At this point in my career, I had written numerous technical bulletins for bomb techs around the world, so I had some respect in the field. I still didn't feel I had accumulated enough respect to tell a bomb tech to his face that his assessment was faulty. But bomb techs are fascinated by the physical effects of explosives. So I knew that if I started pointing out some of the specific effects here with the enigmatic car, these techs would take off on the right road.

In this case, Mark and I had both noticed the pattern of damage to the car. It was obvious that there had not been a bomb in the car. Put a bomb in a car and the force pushes everything outwards. All the damage to the mystery car was from the outside in. The blast had hit from the exterior. If a bomb went off outside the car it had either been in another vehicle or placed by a bomber. The damage to the car was also too intense to have been created by an easily portable bomb. And there was no indication of another vehicle bomb big enough to create what I saw.

Mark and I had independently assessed the scene and had a quick sidebar to confirm our agreement. It was decided I would guide our counterparts through our observations since I was a known entity. I circled the car, bringing our Australian partners around to examine fragment strikes I had noticed. On both sides of the car there were distinct furrows cut along the passenger- and driver-side doors and roof, created from fragments hitting the side and top of the car and continuing down its length. The strikes indicated that the vehicle in question had been facing the bomb. Also, the bomb had been far enough away to ensure that fragments could strike evenly on both sides of the car. The

---

[5] Australian techs were on the scene because Bali is a popular tropical getaway for Australians. They lost eighty-eight of their countrymen in the attack. The FBI was there to assist them and the local police.

damage to the car itself made sense. What did not make sense was the lack of damage to everything in the vicinity of this car.

I came up with a theory I dubbed the "Hand of God" effect. In this case, it was a pure application of Occam's razor (the principle which gives precedence to simplicity: of two competing theories, the simpler explanation is to be preferred). On one hand, the vehicle could have been subject to a yet-to-be discovered bomb. This would require a sizable blast that left no crater, no indication of a blast epicenter, and was placed in front of the car without damaging the car next to it. On the other hand, the vehicle could have originated from somewhere behind the massive vehicle bomb and, as a result of the blast, been tossed in the air only to land between two other cars. It was like the car had been lifted by the Hand of God and set down in a perfect place to baffle forensic bombing specialists.

As my Aussie partners and I puzzled over the origins of the damage done to the car, we did not realize that we'd only scratched the surface of what would turn out to be, well, a weird case.

The Bali attack started in Paddy's Pub, the local version of an Irish pub. Accounts of what happened in Paddy's were contradictory and impossible to reconcile with each other. Only one thing was certain: a bomb had gone off inside the bar.

Paddy's had a unique interior layout, with a central square bar that stood in the middle of the establishment. This rectangular bar sat below a roof-top bar with openings, like skylights, situated above it. The bar was full of customers when the bomb exploded and set off a series of events that resulted in mass chaos. Patrons near the bomb were blown back and peppered with fragmentation. The air blast smashed alcohol bottles, which the flying debris lit on fire. The openings above the bar acted like a giant chimney of sorts, and a raging fire ensued, which made collection of evidence later much more challenging. Somewhat protected in the back of the bar from the initial blast, survivors poured out onto the main street, where the larger vehicle bomb detonated shortly thereafter.

In a typical bombing scene, I work systematically, meticulously combing through the wreckage looking for pieces of the bomb to collect and analyze. Here, our teams had two separate bomb scenes with overlapping areas of damage. Traditionally in such cases, two bombing evidence recoveries would have happened in tandem, run by separate team leaders. However, this was not my country, I was not in the lead, and all I could do was flow with the current and offer whatever assistance I could.

Over the course of the first few days, Mark and I ended up doing more triage than anything else. Our job was to survey the scenes, make assessments of what we thought happened, and confer with the Australian experts to come to mutually agreed-upon conclusions.

One of the unsolved mysteries lurking in the background of this case was the sequence of events that occurred in Paddy's. Witness statements can be inherently unreliable. When the event is as traumatic as an explosion, witness statements are even murkier. Even a question as simple as, "Was it a large blast or a small one?" results in impossible interpretations. For example, flash-bang charges used by SWAT teams to shock and disorient barricaded subjects contain less than a fifth of a pound of explosive. That small amount in near proximity is enough to rattle anyone's senses. It's impossible for a victim to determine how big an explosion is.

There was no shortage of conflicting accounts swirling around what happened in Paddy's. There was one description given of an unknown assailant on a motor scooter who pulled up outside Paddy's and tossed a bag into the window. Another witness described a scenario that included an odd package on the floor exploding. Still another posited it was a suicide bomber who wandered in off the street. As a technical advisor, my job was to go from mystery to mystery and see what analyses I could do to narrow down possibilities.

Sometimes, to solve a conundrum, you have to think outside the box. In the case of Paddy's Pub, that meant resorting to medieval tactics to solving modern-day bombing forensics. To do this, we would need lots of string, tape, and sticks, and enough bravado to sell caveman forensics as sophisticated analysis.

Working scenes like Bali can be dirty and unpleasant. Sifting through the destruction left by a bomb means sorting through human remains.[6] Bombs tear people apart. They produce fragments that rip into flesh and splatter blood everywhere. Bombing scenes are awash in small bits of people. There is no way to keep that off your boots, pants, shirt, and self. We had Tyvek outerwear to try and limit exposure, but the world is full of sharp edges and pointy objects. Bombing scenes ensure that these items are in the majority.

After a few bomb scene investigations, I learned to only use one set of pants and boots. By the end of a long day in the field, my sweat-soaked pants would be covered in grime and miniscule bits of flesh. Every evening, I came back to the hotel and took off my boots outside my room door. I walked into the room and put them on the balcony to disperse the stench of death. Next, I took off my pants and these, too, would go outside. As each day passed, more contamination built up on both my shoes and pants, and

[6] At some point in my professional journey I have lost a bit of my sense of smell, likely due to numerous chemical exposures. There are times I don't mind so much.

the small bits of flesh and goo decomposed just a bit more. Some days, the hardest part of my job was pulling on those slightly stiff, heavily soiled pants and decay-smelling boots for yet another long day in the field.

Every day, one of our assistant legal attachés (ALAT)[7] dropped us off at the perimeter of the crime scene. The Bali streets leading into the epicenter of the blast were lined with bed sheets on which people wrote messages of hope, outrage, sadness, and healing. Every day, more bed-sheet messages appeared. The crime scene area was blocked to vehicle traffic, so we walked a couple of blocks into the bombing site. About half the population in Bali rides motorized bikes and scooters. The other half are in small cars. Both bikes and cars mix and weave in an intricate dance guided by a tempo of horn beeps and revving motors. It is loud on the streets.

Walking in from the cordon tape meant walking away from the traffic. With each passing step you could hear the traffic noticeably retreating into the background.

Eventually the noise of traffic vanished entirely. By the time I approached the bomb site, the still air held a dead quiet. I have never been to a more still and silent crime scene. The feeling was almost one of reverence.

On the periphery of the crime scene staging area, people were allowed to come in with flowers and religious offerings to honor the dead. Small baskets made of woven tropical leaves littered the curbs of sidewalks. Burning sticks of incense peppered the air with a sweet odor mingled with the floral foundation. Bali is a

[7] The FBI has agents distributed in embassies across the world called legal attachés (Legats). They are helped by ALATs. In smaller countries, an ALAT will be the official representative of the FBI in the region and have the job of interfacing with local law enforcement. Legats and ALATs are the most valuable assets we have overseas, in a program I cannot speak highly enough of. Oftentimes the ALAT in Bali would chauffeur us where we needed to go and run interference with local politics to keep us free to do our job.

tropical island—hot and humid. The still air captured odors and seemed to push them into my face and nose.

Had it just been incense and flowers, the memories would be much more harmonious. But bomb scenes bring death. The flesh and blood scattered throughout Paddy's Pub held the stench of decay and the aftermath of the fire. The lingering odor of a campfire combined with that of roadkill almost captures the essence of it. These sights and smells will stay with me for the rest of my life.

At Paddy's Pub, Mark and I decided the most important thing was to figure out where the epicenter of the blast occurred. That's the first clue in determining which of the many stories had the most potential of being credible. A couple of things jumped out immediately. There was a huge mass of viscera and blood on the ceiling. Having seen enough bombing scenes, it was obvious that whoever left the mark on the ceiling had been blown to pieces. However, this could be from either someone wearing the bomb, or some poor soul who had a package tossed through a window land in their lap.[8]

We searched the whole floor and could not find a single sign of cratering. Had a bomb landed on the floor, it would have left an obvious indentation in the concrete when it went off. The odds of someone tossing a package into the window were diminishing more and more as we surveyed the scene.

Multiple support pillars were scattered throughout the bar. These pillars by themselves were nothing fancy, but, as a decorative touch, each pillar had a steel fifty-five-gallon drum wrapped around it. Imagine you're in the room with a bomb and there's a big support pillar nearby. Where would you instinctively go? You'd want to be behind the big-ass pillar for the same reason you

[8] The Australians later took samples from the ceiling and conducted DNA analysis. This came back a little later as being from one person.

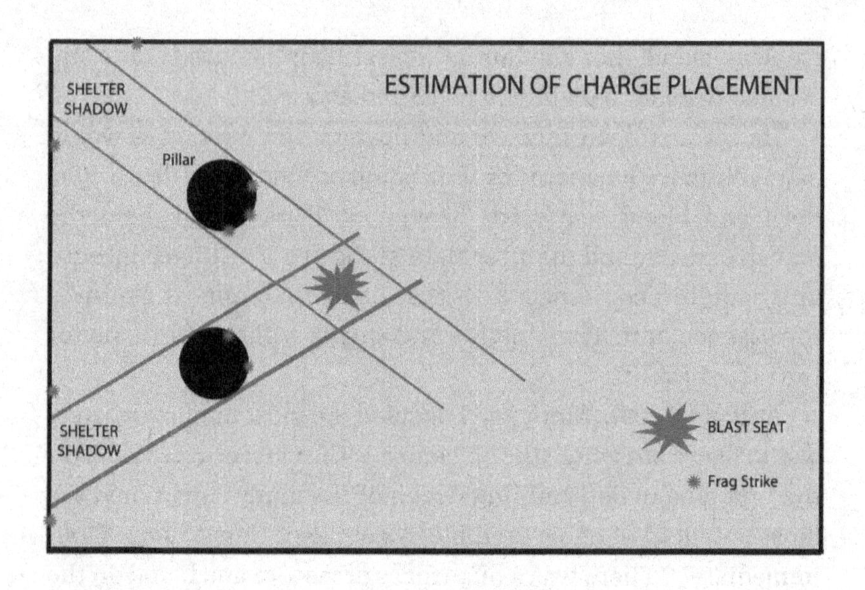

would dive behind it in a gunfight. Fragmentation thrown by an explosion will be blocked by that pillar.

In Paddy's Pub, it was obvious where bomb fragments had struck the fifty-five-gallon drums. It was also obvious where fragments had made contact with the outer walls. If I take a shotgun and shoot at a pillar with a wall behind it some of the pellets will hit the pillar, and some will hit the wall. However, the area of the wall directly behind the pillar will be pristine.

So where do the sticks and strings come into the picture? In Paddy's, I observed a number of pillars and other construction features that would block fragment strikes. The ceiling had uncommonly large beams that came down a foot into the room, creating sheltered areas not only on the walls, but above us, as well. In addition, a couple of sections had long streaks of blood. Imagine wildly swinging a paintbrush soaked with paint. It would leave a streak following the direction of your swing. Mark and I tied off bits of string to the ends of long sticks and ran them in line with the trajectories of blood splatter. We ran other sections of string from beams that had been hit by fragments back to an area that had an open view of every fragment strike.

This analysis went on for hours. By the end of the exercise, we had something that looked like a massive cat's cradle of yarn coming together in the middle of the room. Based on this crude analysis, we figured out that the device was three feet off the ground within a space of about five feet in diameter. With no crater in the concrete, this meant the bomb was sitting on top of a table or was worn, like a backpack or vest. Someone didn't throw it in through the window. Later, it was determined that the bomb was a result of a suicide bomber wearing a vest bomb filled with approximately ten to twenty pounds of TNT.

Of the bombs that went off in Bali the night of the attack, we had ruled out a mysterious fourth bomb, provided some insight into potential investigative avenues for Paddy's bombing, and had peripherally examined the car bombing (more on that in a bit). There was still one bomb that we had almost no visibility on at that point.

At the same time as the suicide bomb and car bomb detonated in one location, a bomb went off against a curb on a deserted street approximately four miles away. The bomb detonated on a road almost directly between the US Consulate and the Australian Embassy. Neither compound was affected by the blast, but both of our teams suspected that our foreign missions might have been targeted.

The street where the bomb went off was still roped off one week later. We had been told the locals had already processed the scene. No one had been hurt by the device, and no obvious target could be discerned. Even with all these facts downplaying the significance of the bomb, both Mark and I decided we should go check out the scene. Our plan was to spend about thirty minutes surveying the damage to get some appreciation for the size of the charge, and at least get some perspective of the area.

We invited along our Aussie partners, who at the time knew little about what the Indonesians had found. Even a week into the attack, the local authorities were being circumspect about what they were willing to share with our teams. As I was an invited guest, it really was not my place to pressure for more visibility. Such matters are left to the State Department.

The street was in a quieter section of town, as embassies are typically located in more upscale neighborhoods. The police tape was still up on both sides of the block. A raised traffic island about three feet high and five feet deep separated this street from a parallel lightly trafficked street. On the other side of the street was the sidewalk.

I did not expect to see much and was ready to do a quick touch and go. It did not take us long to figure out where the bomb went off. In one area, the concrete curbing next to the sidewalk had a divot about eight inches long blown out of it. The gutter was filled with dry leaves, but brushing them aside showed that the divot went down the whole six-inch height of the curbing. The type of damage imparted on steel and concrete can give indications of the type of explosive utilized. My immediate instinct was that the bomb utilized some sort of high explosive.

So how did the bomber set off the device? Did he light a fuse? Was some complicated electronic trigger involved? Did the bomb go off by accident by being dropped from a scooter? Any and all of these theories were possible.

I started looking for other indicators of blast damage. I noticed several trees nearby. One of these trees was pockmarked from having been hit by flying debris. The tree showed deep gouges that I believed could hold part of the answer. I went into my gear bag and pulled out my multitool. Finding a sturdy blade,

I dug into the divot cut into the tree. Within a minute, I pulled out a section of multi-strand wire. A small fragment of wire may not seem like much, but it told all of us bomb guys the same thing: the device had an electrical fuzing system—a power source (most likely a battery) and a switch. We all looked at each other and knew what this meant. We would have to do a serious search of the scene to see what other evidence had been left orphaned.

We were all concentrating so intently on the scene that we failed to notice we were drawing attention. About a dozen locals had parked their scooters on the raised traffic island on the other side of the road and were taking in the free show.

In other countries, the assembly of a crowd might have created unease. But in Bali, where the locals were exceedingly friendly and supportive of our efforts, a dozen scooters could be ignored.

We harvested every hit from every tree in the area, finding more wire and pieces of plastic as we went. Then we rifled through every bush on the other side of the sidewalk to see if anything might have landed in shrubbery. As we made our way back to the road, I noticed that the dozen scooters had multiplied several times, and our audience had grown considerably.

Next, we cleared the leaves away from the curbing next to the sidewalk. This revealed a treasure trove of pieces, including multiple fragments of black plastic that started to look like sections of a cell phone. Sometime during this process, a van pulled up on the other side of the street and a local news crew set up on the raised island, filming our search. We had made the local news now for sure. If our faces weren't known to every bad guy still standing, they sure were going to be now.

We still were finding pieces. Eventually we located a mangled metal square about one and a half inches on each side with a long number stamped into it. I turned to Mark and pointed. "Hell yeah." Numbers are awesome. Through the wonders of the internet, you type in a partial number and, oftentimes, the exact make and model of the electronic item it came from pops up within seconds.

We were almost finished when our peripheral sideshow reached a crescendo. Up pulled an official police van from the local lab. Out of the van rolled two Indonesian investigators in full uniform, with big yellow letters on their jumpsuits, and official gear bags. In an effort to put themselves front and center for the local news, they mimicked our actions.

My team had collected all the evidence that we could reasonably gather considering the circumstances, and—with the situation getting progressively more asinine—we got back in our van and vacated the scene to allow the local crime lab folks to bask in media glory.

We took the pieces we gathered back to command post and within hours the Australians determined they were from a cell phone. They were also able to determine the make and model based on the one fragment containing the stamped number. We had an electrical fuzing system with a cell phone switch.

Having never confronted the Indonesians, I cannot be sure what they collected from the scene. At the time they were not sharing the info. As the FBI did not prosecute this case, my window into the final details is limited. Based on past experience my belief is that the locals cherry-picked evidence from the scene. They most likely found a couple of identifiable pieces of cell phone and wiring and figured what they had was good enough. Obviously, they did not bother to collect everything. Items left behind allowed our team to piece together the same picture, a picture they were reticent to share with us at the time.

Investigations aren't run in a fairy-tale world. Politics, ego, and distrust between countries, even those working on the same side, often interject themselves. Bali was not the perfect case, although things did get better between the various investigators.

The main theory stemming from the evidence was that the third bomber was on his way to the party. For some reason he was not on time. Bombs one and two went off within a very short time span. We think that bomber number three either got cold feet and ditched the device, or that, upon hearing other devices going off, realized that his was next and tossed it to the curb. In either event, it never made it to the location of the main attacks. For that, at least, we can be thankful.

By the time Mark and I arrived at the massive vehicle-bomb scene that created most of the carnage in Bali, the locals had already done much of the heavy lifting in the investigation. The odds of finding small parts of the bomb are nominal in a device weighing a ton. Put a small egg timer on top of two thousand pounds of ANFO and whatever pieces may survive could fly hundreds of yards and end up anywhere. It is the parts of the delivery vehicle that offer the richest evidence.

In both the Oklahoma City and the World Trade Center bombings, the pieces of the Ryder rental trucks recovered from the scenes yielded true investigative clues. Someone rented both of those trucks. In the World Trade Center, remember, the person who rented the truck even came back for his deposit after the truck bomb went off. Seeing how he did not have a truck with him at the time, his request was denied. In Bali, the vehicles had a distinct number associated with them. Like in every major vehicle bomb case, pieces of the truck were recovered. Some of these pieces had that unique number. This number provided investigators enough leads to eventually start tracking down the bombers. Apart from using the truck pieces to develop an estimate of the charge size of the bomb, they did not play much of a role in my analyses.

The chemistry analysis, however, did intrigue me. The Australians brought a full suite of analytical chemistry equipment over to Bali to allow them to conduct real-time analysis around the clock. I had never seen a hotel ballroom transformed into a chem lab before. It was not pretty, but it sure as hell was useful. Early on, they showed me analyses, which showed hits for TNT in Paddy's bar. That made perfect sense. But they also showed signs of the chlorate ion[9] all around the area associated with the vehicle bomb. That made no sense at all.

In the late '90s, I made and studied chlorate-based explosives extensively. While, historically, we have seen bad guys make them in small batches to fill up pipe bombs and maybe small satchel charges, large-scale usage was never attempted. However, the Bali vehicle bomb apparently contained thousands of pounds of it.

In theory, I knew that chlorate could be used to make a large bomb, but history also showed the folly of playing with such materials. The first fully recorded example occurred in 1788 when a French chemist decided he could come up with a better explosive than gunpowder (traditionally a mix of charcoal, Sulfur, and Potassium Nitrate). He swapped out the Potassium Nitrate with Potassium Chlorate. So sure was he of this mixture's potential that he lined up investors for a plant to produce massive amounts of the explosive and decided to proudly show them a trial manufacturing run. The whole party went off for breakfast to give the production run time to start up. Upon their return, they opened the door to view the explosive being made and the plant blew up, killing three of those in attendance. The idea never got off the ground, although several of the investors did. History does not

[9] Some forms of chemical analysis do not give up the whole molecule. Instead they give portions of a compound. Finding the chlorate ion means that some species containing chlorate, the most commonly used in explosives being Potassium Chlorate, might have been present.

relate what happened that fateful day, but many mixtures using chlorate can be sensitive enough to blow up during preparation.

In 1862, two enterprising chemists decided they figured out where their colleagues had gone wrong in 1788. They mixed Potassium Chlorate, Sulfur, and several other ingredients[10] to come up with something they optimistically named "Safety Powder." In full defiance of Murphy's Law, they called their method of making this new material the "Anti-Explosion" process. Between 1862 and 1865, their method resulted in five explosions during production. The last detonation was so massive it demolished an entire city block's worth of buildings. Thankfully, this idea never took off.

Perhaps the best warning about chlorate explosives was penned by the godfather of all explosives, Alfred Nobel himself. Seeing the disasters befalling scientists working with these materials, he sent a letter of either warning or encouragement. In part it read, "Your fear of chlorate of potassium is exaggerated. When it smells of sulphur it is as sensitive as a hysterical girl, and when it feels phosphorus on its surface it is worse than a thousand devils. But it can very well be tamed down to keep itself within the nurture and admonition of the Lord." This remains one of my all-time favorite historical quotations.

What Nobel was trying to say in layman's terms was, "Dumbasses. Don't mix Potassium Chlorate with Sulfur or Phosphorous." The explosive charge in Bali, shockingly, did just that.

Over the course of the investigation, we learned that the bomb employed a main charge of Potassium Chlorate mixed with Sulfur—just like the formulations that killed two rounds of inventors centuries ago—and Aluminum powder. Sulfur makes mixtures sensitive to friction, and Aluminum makes mixtures sensitive to static. Chlorate and Aluminum mixes have been

[10] I hate to be vague. But for reasons of safety and security, it's for the best.

known to be set off by the static spark people generate shuffling across a synthetic carpet on a dry day. The Chlorate/Sulfur/Aluminum chemical pairing matches what was historically found in cheap knock off pyrotechnics. In the '60s, a firecracker referred to as an M-80 contained approximately three grams of a very similar mixture. M-80 rose to fame for its inclination to blow off the fingers of children who held them a little too long. They were made illegal over fifty years ago.

The Bali bomb packed two thousand pounds of a similar mixture into a truck. The magnitude of this risk stuns me to this day. Most large vehicle bombs across the globe are based on different types of fertilizers. Ammonium Nitrate (AN) is one of the top choices. The terrorist cell in Bali could not get AN. There was also no way they could get their hands on tons of military explosives like TNT. Geographically speaking, their highly volatile Chlorate/Sulfur/Aluminum mixture was quite accessible to this terrorist cell. Indonesia is in close proximity to China, the world's top producer of both fireworks and the chemicals required to make fireworks. Bad guys are always willing to take risks. And sometimes their lack of choices forces them to proceed at very high stakes. Cases like this clearly indicate how desperate they are to deliver their deadly message.

Bombings can devastate a country, and, by the very nature of my job, I see people after they have been through some of the worst of life's horrors. My presence on the scene of a bombing lasts only a short span of time. In overseas deployments I rarely get to see the whole story, often don't know the full value (if any) of my contributions, and—when all is said and done—am never sure how the tale ends.

Following the bombings in Bali, the tourists all vanished. My team was put up at a four-star hotel (for well under a hundred dollars a night) for the duration of our investigation. The entire hotel was deserted and sat in creepy silence. It would remain so throughout my stay. The beaches and local markets, too, were devoid of people. It was as if the world had deserted Bali. And the Balinese felt it.

Approximately 87 percent of the population of Indonesia are practitioners of Islam. Bali is unique in that over 80 percent of its populace practices a special form of Hinduism referred to as Agama Hindu Dharma. Part of the philosophy of this belief system is Tri Hita Karana (Three Causes to Prosperity). Greatly simplified, the three principles are harmony among people, harmony with nature, and harmony with God. I found the Balinese people to be very giving and peace-loving. The atrocity of the bombing was anathema to everything they believed brought meaning to life. I have done my best to limit my impressions of Bali, as I did not see it at its best. But there's one encounter that will stay with me for the rest of my life. I had a free hour a day or so before I was scheduled to fly back to the States. I decided to head to the local market pick up some trinkets for my wife and then-young boys. The typically bustling thoroughfare was stall upon stall of merchants sitting in silence. I had brought along another FBI team member who had arrived earlier in our deployment, who had been deployed to serve as a translator for us. It is never clear when you are going to pick up a life lesson. But she taught me one that day.

I was looking through shirts in one stall and asked, through her, what the price was. I forget how many rupiahs it cost. Basically the cost per shirt came to about four dollars US. This young lady then engaged in an energetic haggling session with the merchant. She saw I was about to interject, so she turned to me and stopped me mid-stream. She said, "I am your wife right now, and

I control the money." Without missing a beat, she went back to the vigorous negotiations. I think she saved me a dollar a shirt.

As we were walking away, I asked her what the hell that was about; I would have given the vendor twice the asking price. Basically, I was thinking like an American and feeling sorry for the hardship brought on the people. The young lady was thinking like a local and giving the vendor the respect of bartering for his wares. I was trying to give him money; she was giving him dignity.

Sitting in the empty hotel bar late one night, I struck up a conversation with the bartender. At one point he asked, "Do you think Bali will ever be normal again? Will the tourists want to return after this?" sweeping his arm towards the seemingly insurmountable wreckage.

I paused, weighing my words to offer hope while acknowledging reality.

"Terrorist attacks are horrific and it's all going to feel overwhelming for some time," I started. "But with every attack, good people rise up and help each other. They pick up the pieces. So yeah. The wound does heal. Things never return to exactly the way they were before. There's an innocence that is lost. But you'll rebuild and move forward. Bali will recover." The bartender looked a little sad and just said, "I hope so." I did, too.

The Bali bombings raised a great deal of questions in the explosive community. We had never before seen a two-thousand-pound bomb made of a flash powder mixture of Potassium Chlorate, Sulfur, and Aluminum. When a new explosive hits the scene, I typically get lots of inquiries about what I know about the material. I knew the explosive sensitivity of the mixture to a fair approximation. However, I had no clue how powerful it might be at the ton level.

Explosives are very complicated things. A mixture might simply catch on fire and burn when in small batches, such as a shot glass worth. Put that same mixture in a cardboard tube and the hot gases it produces upon burning are trapped. This makes the reaction faster. Now that same amount of material that caught on fire in the open might explode the tube. If a pint glass of mixture is made, it may no longer burn. The weight of the powder itself might provide enough confinement to produce an explosion. In Bali, there was not a shot glass of flash powder. There was a ton of it. No one had ever studied it in quantities that large.

When I returned to the lab, I immediately put together a research proposal to conduct a detailed research program on flash powder. The FBI granted me money to move forward with this study.

There was no way we could do the research at the FBI. To start, the FBI explosive ranges can only handle fifty-pound charges. Put thousands of pounds on our ranges and windows would be blown out across the FBI complex a short distance away. In addition, the flash powder mixture was horrifically sensitive to friction and static. There was no way anyone could hand-mix hundreds of pounds of this explosive. This work had to be contracted out.

Only a small handful of specialists make and study terrorist explosives. When we put out a request for people to conduct this dangerous research, we received limited takers. In the end, we worked with a company employing scientists I had helped train years before in New Mexico. The company had some very clever engineers in its ranks who designed a fully remote-controlled process to mix and move the explosive. The flash powder was made in fifty-pound batches. Scientists loaded the ingredients into two separate hoppers (Potassium Chlorate in one, Sulfur and Aluminum in the other). As long as the Chlorate is not mixed, it is totally safe. Once the hoppers were filled, everyone retreated to a buried bunker, and the dance began. The techs remotely opened the hoppers to discharge their contents into a cement mixer. The mixer was fitted with a special static-proof bag on the inside. Once the mixer had all the ingredients added, the techs remotely turned it on to allow the materials to tumble together.

Once mixing was finished, the fifty-pound batch of irritable explosive had to be transported. The team's electronics whiz put together one of the coolest toys I have ever seen on a range, intended to transport the sack of death. The engineer came up with a moving A-frame gantry with attached claw. Imagine the crooked claw game at a carnival that grabs at stuffed animals.

Now imagine the same claw ten times bigger and actually functional. Both the claw and the A-frame were controlled by a model airplane control system. The gantry had wheels powdered by multiple car batteries. (The picture shows the apparatus without the claw attached.) After the charge was mixed, this mechanism rolled in, grabbed the bag from the mixer with the claw, and carried it down to the testing pad. Bags were heaped in a pile until the charge weight needed for the test was achieved.

Early on in testing, we had many safety discussions. One question that arose was, "What would we do if the robot broke down or dropped a bag on the ground?" No one was going to head down range and try to pry a fifty-pound sack of hypersensitive explosive from a recalcitrant robot. Remote disposal was about the only answer: to keep from risking someone's safety by having them approach the mixture, we decided to test a remote method to dispose of the mixture. A fifty-pound sack of the flash powder mixture was prepared and placed on the ground. From two hundred yards away, a range technician shot the bag with a high-powered rifle. The mixture detonated instantly, and we had our remote destruction protocol.

Testing took almost a year to complete, but in the end we learned a great deal about flash powder mixing. We learned that flash powder may not explode with the power of TNT when it is placed in a small firecracker, but once a hundred pounds are set off, things get much more serious. There were also various tricks we learned to getting optimum energy from the mix, which will not be shared here. The Bali bombers appeared to have applied some of these techniques, which was another valuable insight.

On the flip side, we also learned some of the things the mixture was not capable of doing. Unfortunately, following the Bali attack, the mixture gained an almost mystical reputation. Intelligence specialists without any real understanding of explosive chemistry wrote assessments surrounding the tremendous power and devastating effects of the mix. One attribute in particular started to develop its own lore. This was the thermal effect of the charge.

Crazy speculation emerged about how the Bali bomb was specially designed to produce a massive thermal pulse capable of much greater devastation through fire than TNT. The reason for this speculation was the fact that so many of the victims of the Bali attack showed significant thermal burns. Such burns are not as common with other explosive attacks. Those who made this leap in logic did not understand the difference between correlation and causation.

Correlation is a relationship or connection between two things. In this case, there was a correlation between the Bali bombing attack and victims experiencing a higher incidence of thermal injury. *Causation* is when one thing has a direct attributable effect on another. In Bali, the bomb itself did not cause the burns. Most of the burns occurred in the nightclub, which was the main target of the vehicle bomb. The nightclub was open-aired with a tiki-bar feel. The main bar was covered with a thick grass thatch roofing material. When the bomb went off, all this thatch collapsed and caught on fire. The fire from this material was the cause of the burns, not the explosive used in the attack.

Part of my research was to measure the thermal pulse and fireball temperature of the flash powder used in Bali. At the same time, we also used charges of equal size TNT. What we found was that five hundred pounds of TNT actually put out more thermal energy than two thousand pounds of flash powder. The lore of this charge's thermal effects lives on, but the science tells another story.

CHAPTER 4

Reconstructing Chaos Within Chaos

In the early 2000s, a car bomb detonated near a shopping center in a South American city.[11] The center contained the expected movie theaters, restaurants, and shops. The bomb, which was in a car parked directly in front of these local business, devastated them, damaged nine cars in close proximity, and blew out windows in a nearby hotel. As a result of the bombing, scores of people were killed and injured. Amongst the dead was a young man buried in rubble and another man who had been thrown 160 feet by the blast, his leg now missing. It was our job to try to find what caused this tragic separation.

When a car bomb goes off, it sends mangled remnants everywhere, from oven-size shrapnel to itsy-bitsy slivers of wire to

[11] I would love to tell you the city and country, but some cooperative agreements make that challenging. Fear not, for the story is about the response and lessons learned from it. Unlike in real estate, location doesn't matter here.

fine residue that coats everything in sight like the lightest layer of silvery new-fallen snow. It's all evidence. And it all needs to be analyzed.

This extremely tedious work often involves rubbing a sterile cotton swab held by a pair of tweezers over walls and chairs and street signs, and anything else close to the bomb, so you can run what it collected through instruments that analyze it. That's not a simple task even when you have full, easy access to the scene and are in a country where pretty much everyone speaks your language and is glad you're on the job.

Now imagine trying to do the orderly function of a forensic scientist in a place where everyone's screaming at you in a foreign language and you have no idea if they're screaming at you, about you, or to you...and they don't always seem particularly happy that you're there.

This is all too often our reality. Nowhere was that point driven home with such crystal clarity as in South America, where I was deployed to assist in evidence recovery and field examination following the car bombing in the shopping center. It was my very first international deployment, which was akin to being tossed into the middle of a quarry wearing one cement shoe.

We arrived at the scene of the crime to find...no scene. No shrapnel. No mangled metal or shreds of tire treads. Nothing. Yeah. We were a bit surprised that a car could explode without a trace, too.

My training agent Tom Mohnal[12] (the FBI agent lucky enough to be assigned a scientist—me—to teach the way of FBI investigations) and I were among the first US examiners to arrive after the bomb went off. The United States wasn't the target. The bombers hit a business that their group had a grievance against. But that

[12] Sadly, Tom passed away during the clearance process for this book of complications from COVID. With Tom passed some of the greatest stories in the world of bombing forensics.

business happened to be across the street from a US company, so a whole bunch of car parts and debris flew into US hands. That suddenly made it our problem. But that wasn't our only problem. A high-ranking US official was due to arrive in a few days for a State Department visit. Any time any high-profile US official travels abroad there is always the paranoia that bombers will be gunning for them. The State Department wanted to make sure the first bomb wasn't a practice run.

Totally ignorant of the politics at the time, we were flown in to figure out what the hell was going on, a task that proved to become more challenging at every turn. And it started with the bomb scene that wasn't a scene. It was literally the most pristine bombing site I'd ever witnessed.

The local government, in an effort to present the US diplomat with a more aesthetic view of a mall that had to be driven past on the way to the US Embassy, had swept up the street, scooping up every car that was damaged and piling them in a huge heap in the alleyway next to the police station. Instead of taking in the usual clues from a scene, we ended up just staring at this towering pile of crap that resembled the aftermath of a *Mad Max* crash scene.

Sadly, neither car bombs nor bombing campaigns are new to many South American countries. Probably the best historical example I can draw from would be the experiences of Peru. At the time of my deployment, the last car bomb to terrorize Lima occurred in May 1997 and was the work of the Shining Path insurgency. The Shining Path, or Sendero Luminoso, was a revolutionary group that emerged in Peru in 1980. Founded by Abimael Guzmán, a big fan of Maoist communism, it sought to create a peasants' revolt to overthrow the government. Some thirty thousand people died in the violence of the 1980s and 1990s during the insurgencies of the Shining Path and the smaller and less deadly Túpac Amaru Revolutionary Movement.

The Shining Path bombers' main tool was dynamite stolen from small mining camps that proliferated across the rural Peruvian landscape. In 1986, Shining Path guerillas stole ninety-two thousand sticks of dynamite in one raid alone. This dynamite was utilized to bring terror to Lima throughout the '80s and early '90s. Shining Path tactics were well-coordinated and highly effective. In one series of attacks in 1983, guerillas bombed ten electrical towers and threw Lima into total darkness. In the confusion, they then proceeded to set off between thirty and fifty bombs at a wide variety of government and capitalist targets. The US Embassy was targeted by multiple dynamite bombings during this time period.

The arrest of Guzmán in 1992 led to a slow dismantling of the organization. Eventually, tensions eased, and bombings abated. The citizens of Lima put this chapter of their history behind them, began to relax, and looked forward to a more peaceful future.

In many ways, Peru's "peasants' revolt" against the government echoed the anarchist fight seen in the United States during the turn of the twentieth century. At that time, workers in the States turned to dynamite to fight against the capitalist robber barons, who built their fortunes upon the backs of the working class. Both peasants and factory workers sought out the great equalizing force of explosives to aid in their cause—a tale that repeated and continues to repeat itself in countless conflicts, both real and perceived.

Peru is not unique in enduring such attacks. Colombia faces one of the more widely recognized terrorist groups in the Revolutionary Armed Forces of Colombia, or in the native language Fuerzas Armadas Revolucionarias de Colombia (FARC). Funded by cocaine trafficking, the FARC has ample resources to produce a wide variety of bombs. Bogotá faced multiple bombings over the years. In 1989, a van laden with approximately one thousand pounds of dynamite detonated near Colombia's police intelligence headquarters. An initial tally listed forty-five dead and four hundred injured.

If you are in the business long enough you get to see the cycle of chaos, cessation of hostilities, and reemergence of violence. In November 2016 the Colombian government signed a peace accord with the FARC. This marked the end of five decades of violence. Colombia may have cooled the conflict with the FARC, but in 2019 the National Liberation Army (ELN) would bring Bogotá back into the crosshairs. A car bomb attack against the National Police Academy would kill twenty-two police cadets.

Colombia, Brazil, Peru, Argentina, and many other South American nations were no strangers to urban bombing attacks from a diverse array of terrorist groups. My first deployment would be my introduction to their aftermath.

I had been in the FBI for a short time at this point. I had worked on various searches on domestic cases and examined a good deal of evidence. I knew explosives and had a fairly decent grasp of how to conduct forensics on bombing evidence. But I had never played this game outside my home stadium, so to speak. In fact, I had never really traveled outside the US. I grew up in a small town in Pennsylvania, with a population of about six thousand. My dad was a schoolteacher; my mom worked the front desk at the local hospital. The first time I was on a plane was during my junior year in college. I was far from a seasoned world traveler.

I imagined an FBI deployment akin to serious men in dark suits sitting around big oak tables, eyes fixed on monitors showing the situation on the ground, preparing to launch the crack team out on the Black Hawk. My image of what deployment to a bombing scene from the FBI looked like was far from the reality.

Most FBI missions to render assistance have a political undertone. In many cases, the attacked country requests FBI help because it does not want to be seen as incompetent to its people. Some, begrudgingly, have to acknowledge they are out of their depth and they need some technical assistance. The FBI typically sends a small team to render this assistance, even if the aided country has no idea what type of help it needs.

The real-life unfolding of actions is far removed from the scenes you watch in movies. Our manager called down to the FBI Lab. "Hey, something just blew up. Can you send some bomb guys out to help the locals?" Armed with that treasure trove of information, the team grabbed our gear and prepared to tackle the unknown.

I left my home state of Virginia in the spring and landed in what felt like tropical August in South America. We were met by an entourage from the embassy and whisked away to our lodging. It was after one in the morning when we headed out of the airport into bumper-to-bumper traffic and the overwhelming smell of petrol and diesel fumes.

Having spent most of my formative years in a small town, college years locked away studying, and first part of my professional life in an even smaller town in the middle of the desert, I was not prepared for the activity of a major South American city. I kept wondering where in God's name all these people were going at one in the morning. After a few short hours of sleep, we woke up early the next day and eked our way through more bumper-to-bumper traffic to the embassy. At this point, the team still had no idea what our mission was to be. Inching along the highway, I talked to Tom about what we should do once on site. Tom, the forensic examiner who had analyzed all the Unabomber devices, possessed a wealth of knowledge. The plan he laid out would become a battle rhythm that defined much of my career for the next decade.

As a scientist, my main concern is collecting and documenting evidence. At every bomb scene, two things hold the most interest to me. The first is the bomb. I can't say anything about the device unless I see either it, or the pieces that are left after it goes off. The second thing is the scene of the crime, where the bomb went off. The damage done to the surroundings following an explosion can reveal subtle clues if you are trained to read them. There is no school that teaches this skill set. I had a PhD in chemistry when I joined the FBI. Other than teaching me how to think like a scientist, that degree meant very little in my day job. What really mattered was that I spent nearly seven years prior to joining the FBI making explosives and bombs and blowing them up. I had seen the carnage of high explosives visited on hundreds of cars. I had witnessed the interplay between shockwaves in air and a wide variety of barriers that unwittingly got in their way. After many years blowing shit up, I had developed the ability to see patterns in apparent chaos.

The embassy staff, some of whom responded to the scene, were kind enough to collect some evidence and hold it for us. It may sound like a small thing, but having access to anything from a bombing scene that is guaranteed to not have been touched by a multitude of curious and unwitting people is a blessing. I knew we had a decent chance of getting some explosive residue from the items the embassy collected.

Residue is the most fragile form of evidence. It is also the form most likely to be ruined by contamination. When the bombing occurred, it started numerous fires around the seat of the blast. You can't fault folks for putting out fires, but some explosive residue is water-soluble, and fire hoses aimed at anything that does not move within a blast site obliterate these types of residue. Even if the residue is not water-soluble, it consists of microscopic particles of explosive that hitch a ride on a water plume straight down the nearest sewer grate. We got to the scene days after the event. In that time, fires had been fought, and the scene had been trampled by police and military personnel and opened back up to the public. All hope of finding post-blast residue was a long-faded dream. Conversely, the car parts collected by the embassy were collected right after the blast by trained embassy personnel wearing gloves and brought in out of the weather. This evidence actually had a decent chance of yielding forensic clues.

In a major bombing, we might take the evidence back with us. However, this case was pretty clearly a local concern. For our own intelligence, we took swabs from the evidence. Swabbing is a fancy term for swiping a surface with material in hopes of removing the microscopic particles of explosives. There are many ways to do this. In the FBI, we typically scour the surface of interest with sterile cotton balls. These are put in a sterile glass jar, which in turn goes into a sterile paint can, and gets shipped back to the lab.

We go to great lengths to prove the explosive particles did not hitch a ride along with us. All material we use to collect residue is first tested to show it is clean. Once on scene, we collect control swabs. For example, before I swabbed the car parts I put on sterile gloves (again tested in the lab for cleanliness). I then swabbed my gloved hands to prove I was not the source of any explosives that might be found. I swabbed the room the car parts had been stored in to prove it was not full of explosive residue.

Such meticulous attention to process and detail is necessary. If, for some reason, this case, or some offshoot of it, came to US court, the evidence has to stand scrutiny above and beyond what many countries would apply. As this was largely deemed a local matter, it was good to get field experience on a case that most likely would not come under that type of scrutiny.

Having collected residue from the embassy evidence, we were then escorted to the bomb site. Part of me hoped to be delivered to a plaza cordoned off with yellow police tape and a big hole somewhere past that tape where an explosion occurred. What I did not expect to see was a shopping mall that looked like it had just finished its ribbon-cutting ceremony. Cognizant of the eyesore that broken windows, shattered shop fronts, craters, burnt-out cars, and scattered body parts would create as the backdrop to the US diplomatic drive by, local authorities diligently went about the job of totally refurbishing the entire scene.

Of the two main interests I had, one was completely gone. Without a crime scene, there was no crater to measure, no window breakage to analyze, no structural damage as a function of distance to observe, and no reason to be standing around a shopping center that used to be a bombing scene.

But then the embassy staff and local police said they could show us the vehicles. They noted that all the damaged cars had been taken to the police station nearby. A small flame of hope built up inside me. If they had actually gotten the vehicles into police custody and put them in a police lot, I might have some data to collect after all. Perhaps something in the conversation had gotten lost in translation. The only part of the story that turned out to be true was that the cars had been moved to a nearby police station. However, instead of some well-organized collection of vehicles from the scene arranged in a lot, we were greeted by a pile of metal in an alleyway. It seemed any cars that still contained a glimmer of hope of being fixed up had been taken off for repair. Any mangled cars left were heaped on top of each other like a short stack of pancakes.

It is protocol for us to take photos of the cars to record the damage suffered and measure how far they were from each other to estimate charge sizes and sometimes orientation of the device.

But when all the cars are loaded onto a truck and dumped into a mass grave, the measurements lose their utility.

Tom, being a "car guy" and our most seasoned bombing investigator at the time, decided to see if he could get some VIN numbers off the bomb vehicle. Most people know what a VIN number is; what many people don't realize is that there are more of those numbers hidden around a car than you could ever guess. At this point in the investigation we knew very little about the attack, as nothing had been shared with us by the local law enforcement. So, Tom decided to play king of the hill and ascend the mound of rickety car parts. It was not hard to figure out what parts belonged to the bomb car, as they were the smallest pieces.

He eventually did find the VIN for the car, and we both came away feeling like we had done some valuable forensic investigation. However, a day later we discovered local authorities already knew exactly where the bomb car came from. It was a hard lesson I learned on this first outing: don't count on being told anything, and when told something, don't count on it being right.

In charge of every investigation there seems to be a general that needs to be met with. Our next stop would be the "dance of dignitaries." There are some international rituals that, to this day, I just don't understand. South America was my first introduction to the Opening of the Coke Bottles. In all the impoverished, hot, and humid countries I visited in times of chaos, meetings with dignitaries always started with a wait staff member bringing in a tray with miniature Coke bottles, just like I remember from vending machines in my childhood. The warm afternoon air wafted in through the open windows of the general's large office. Coke was cracked open and poured into small glasses, with all due solemnity. You knew you were really being granted VIP status if ice cubes were part of the bargain. So there I was sitting in a general's office, sipping my slightly-cooler-than-room-temperature Coke when the next ritual began: the Face-Save Shuffle.

Typically, when we work one-on-one with our international counterparts, they are grateful for everything we can do to help. The higher you climb the executive ladder, however, the more our presence can sometimes be perceived as a sign of inherent weakness. The local authorities sincerely expressed how they had solved the case and were eager to share this knowledge with the FBI as a learning experience for the Bureau. I have learned over my career to smile and thank these types of people, but in South America, my first international experience, I couldn't help thinking, "Dude...I mean General Dude...you have a big-ass heap of cars rusting in an alleyway that were carted away from a scene that was bulldozed practically before it was searched." None of this was vocalized.

At some point in the conversation the general stated that they were almost certain they knew what type of bomb was used in the attack. He offered to let us see what he was talking about. At this point I expected the general to pull out a folder with interview notes and crime scene photos. But what he did next was so outside the realm of what I considered possible that at first it failed to register.

The general got up from his chair and went into what looked like a miniature locker room just off his main lounge area. He came back out carrying a very large shoulder-slung gym bag. It was bigger than the suitcase I carry with me on weeklong trips. He sat it down on the table in front of our team. We asked what it was, and he explained it was a bomb that they had interdicted a couple days before the successful bombing.

Only two of us in the room were explosive specialists. Tom was a bomb tech, and I had spent years replicating bombs. We both turned and stared at each other for what seemed like a socially unacceptable amount of time.

The general invited us to look inside the bag. On this trip, I was what is referred to in less-than-civil society as the FNG (Fuckin' New Guy), so the task fell on me. I opened the zipper and revealed what appeared to be a shit ton of ANFO. A powerful high explosive with a long history in bomb making, it does not have any business hanging out in a government building stored away in a gym bag in a shower room.

The comforting fact—if there is such a thing as a comforting fact when dealing with terrorist explosives—was that ANFO is not easily initiated. It takes another hunk of explosive put into the material to set it off. However, the thirty to fifty pounds of material sitting in front of me would be enough to destroy this section of the building and ensure that everyone in the room would need a closed casket at the funeral to follow, provided they found enough of us to bury.

The voice in the back of my head was not panicked, but rather insistent in a *Rain Man*-like fashion: "This bomb doesn't belong in a building. This bomb really does not belong in this building. This is not a good building to store a bomb."

Suddenly, I noticed a sphere about the size of a softball wrapped in black plastic garbage bag and circled with multiple loops of twine. Rain Man shut up and a more urgent voice took his place: "Danger, Will Robinson. Danger."

On its own, ANFO is safe; it needs a hunk of explosive called a booster to set it off. But this sphere.... I was looking at something the exact size and shape I would expect for a booster. And that hunk of explosive is easier to set off than the ANFO, say, possibly with rough handling by some fool opening a bomb bag in a government building.

I looked up and saw that Tom had taken a step or two away. This is a purely reflex reaction. With an explosive that size, he would have been just as dead as me. I had to get that sphere out of the main charge.

I asked if I could take some samples of the material for the FBI. I opened a sampling kit I had brought along with vials the size of mini shot glasses. The local authorities had pounds. I wanted a lot less than pounds. Seeing the vials, they agreed. This gave me the excuse to gingerly pull the Black Ball O' Death out of the ANFO and set it to the side. Now, at least, if something went wrong it would only be me flying home in the cargo hold. With my multitool, I started to cut away twine and unwrap the plastic baggie covering the ball. Then I froze.

Behind some of the plastic I spied two small holes about an eighth of an inch in diameter. Boosters don't just go off by themselves. They need a little kick in the ass to encourage them. This impetus comes from a detonator, which would be stuck into the booster in holes just the size I was staring at. I had not one but two locations where the next part of the bomb could be hiding. It seemed a good time to ask some follow-up questions.

I turned to the embassy team and politely requested that they inquire whether anyone had x-rayed this device. I am not superman, and I could not see into the sphere I was holding. An X-ray could, however, see if two small metal tubes were buried in the hunk of explosive I was holding. After about a minute of spirited discussion in Spanish, the embassy translator said simply, "It's fine." I really didn't want to be a dick on my first time out of the country, but I felt it important I clarify myself in case something had been lost in translation. I asked again if any bomb techs had x-rayed the device. Again, conversation went back and forth between the assembled folks. This time I was give the response, "Everything is safe."

There are phrases that don't engender the sentiment they are meant to convey. Amongst these are "Trust me," "It's fine," "No worries," and "Everything is safe." These phrases ring with an even more hollow tone when you are in a foreign land holding a couple pounds of what appears to be explosive of unknown origin, which may or may not have a detonator buried within its depths. A smarter man might have put down the bomb and politely declined to examine it any further.

In the tarot deck, there is a card referred to as The Fool. It is a picture of a man stepping off a cliff's edge. One foot is planted solidly on the ground while the other hovers over the space marking the start of the abyss. With my foot poised in midair, I switched to a scalpel to take off the last part of the plastic covering the sphere. Dynamite! The main ingredient in most dynamites is the highly unstable liquid explosive Nitroglycerin (NG). Dynamite is a mixture of NG absorbed on something designed to make it hate life just a little less and persuade it to stick around a little while longer. I was holding onto a two-pound ball of dynamite. On the bright side I could see the holes were empty. While the risk of an explosion had diminished, I was not out of the woods.

Muscles still tense, I focused on sampling the ANFO and dynamite. Sampling the ANFO was easy. I quickly scooped a portion of the material into the vial. Sampling the dynamite was a little trickier. As a general rule, dynamite of unknown origin is always treated with high caution. With the smallest blade on my multitool, I planned to gently cut out a pea-size divot.

I had just opened up the blade when the room erupted into chaos. People were shouting in Spanish, frantic for us to leave. Having come close to being startled to the point of dropping a two-pound lump of dynamite in my lap, I found myself suddenly bereft of diplomatic tact. I could not go anywhere fast as I was holding a bomb component in one hand with a knife stuck into it with the other. With the chaos swirling around me, I grabbed my sample. My teammates were pushed to the door, but no one in the room was willing to wrestle with the man holding the ball of high explosives and the knife. The minute I put the explosive down, I was grabbed and herded out with my sample kit in hand. The team was whisked away to a small corner room at the end of the hall and the door shut behind us.

For fifteen minutes, we sat, bewildered, at what had happened. Did I just create a diplomatic meltdown on my first time out of the country? Finally, the team was escorted away, and we were clued in as to what happened.

Turns out, the general's office we had been in was on the same floor as a high-ranking government official who was his boss. As I dissected the dynamite, one of the general's men noticed the official's motorcade pull up. As the official was unaware that the general had a complete bomb stuffed in his locker room, the general felt it was the best thing for his continued career if his boss maintained the same level of ignorance surrounding the big-ass explosive charge parked nearby. Even better if the boss did not see an American FBI investigator cutting into a huge wad of dynamite as he passed by.

So what did I learn from this first jaunt out of the country? To this day, I am still not sure I can condense it down to a simple synopsis. I realized that I had no way of knowing what future chaos I would encounter at bomb scenes. I went in with no idea what I would be asked to do. I hit the ground with no ability to do what I thought I should do. I left with no idea what I had done. Tomorrow, I could wake up and find myself doing the same thing in another country. So, I learned to arrive with no expectations, to help wherever able, to keep a poker face as best I can, and to breathe a sigh of relief when the plane is wheels down back in the United States.

One of the things I never anticipated was how little I would ever get to know about many of the cases I worked on deployment. In many instances, the local military or law enforcement just does not feel enough trust to share what they know. In others, the locals just don't have the information you need. I can never tell if the lack of information I am getting is by intent or ignorance. In some cases, I hear odd rumbles of a story years later from people who were on scene and heard bits and pieces from their own reliable sources. South America was just such a case.

In full disclosure I do not know the whole truth behind this next bit, but I have reason to trust my source. The locals were nervous about the bomb that went off for a very particular reason. The week before the shopping-center bombing, the police had been approached by a man who claimed he was given a bomb by the bad guys and told to blow something up. He lost his nerve and ran to the cops. The police gave him a small amount of explosive back and told him to place it in a target they selected—in this case, a deserted telephone booth. With only a fraction of the explosive being used, the damage was of no consequence. The man came back to the police and told them the terrorist group was not satisfied and they wanted to give him another bomb to conduct another attack. Police told him to stall. Very shortly after this, the bomb went off in the shopping center. Police believed the same group was responsible for both attacks. As it turns out, the bomb I took apart in the general's office was the very same bomb the police got a hold of in the first place. The story makes sense out of why authorities had such a full-up device.

Another thing I learned while in South America is that there are a huge number of executives back at the mother ship who take a keen interest in the folks on the ground investigating a bombing. Unfortunately, these executives really don't understand enough about bombings to know what pertinent questions to ask to sate their curiosity. But above all other questions, the one I hear the most, which I was first peppered with continuously in South America, is, "How big was the bomb?"

Size seems to be a fixation in numerous areas of human endeavor. In South America, answering the question of size was made monumentally difficult by the fact that there was no crime scene to examine. There was only one data point that I had to fall back on. One of the embassy security staff who deployed to the shopping plaza following the bombing took a distance measurement of one of the damaged cars. I have no idea why, but he did. As a result, we knew that a car parked in the spot right next to the bomb vehicle was thrown seventy-five feet.

One day, in a sarcastic response to another query to estimate the size of the bomb, I stated that the only facts I had to go by were that the main charge was most likely ANFO, and a car parked next to the bomb got tossed seventy-five feet. Then I proclaimed that when I got back to the lab, I would build a whole bunch of ANFO bombs to see how far I could throw cars. My sarcastic retort was met with resounding enthusiasm. Thus, Tom and I found ourselves spending a few days on our demolition range blowing the living crap out of vehicles.

I want to stress that I am a scientist. There are many ways to do science. There is a methodical approach, with control of every variable possible combined with meticulous measurements. Then there is the *MythBusters* extreme approach, which is seeing how much damage you can do with explosive charges. I love *MythBusters*; I watched every season of the show with my boys when they were young. But they are *not* doing research. They are doing rough proof-of-concept tests. My vehicle demolition tests were more in line with their approach.

We purchased vehicles that were the same general size as the bomb vehicle and target vehicle. We used the same explosive. Every explosive gives off its energy a little differently. ANFO is known for a little slower energy release. In the field, we say it tends to heave things. So, putting a fifty-pound charge of C4 next to a car would rip it to shreds. The same amount of ANFO will not tear it into as many pieces but will instead produce a prolonged push against it that will displace it further.

The issue was trying to simulate the same ground surface. Try bouncing a tennis ball off of concrete, and then bounce it off a thick, grassy lawn. When an explosive goes off, the pressure hits the ground and bounces back. If the ground is soft, more energy is absorbed and taken up, creating a crater. The bomb went off on a city street, which allowed the pressure to reflect upwards and against the vehicle parked next to it. The range was hard-packed dirt, which is more giving than asphalt. To compensate for this fact, we brought in concrete pavers to place underneath our bomb vehicle to create more reflection. We also placed the bomb in the same location witnesses claimed it was placed in the South American car bomb.

With all the variables we could control in place we set off blowing things up and taking measurements. This may sound like fun, and it is at first. But it takes hours to clean up a car after an explosive charge distributes it across a range. So for every few seconds of thrill, there are hours of monotonous clean up.

In the end, I built ANFO charges of twenty-five-pound, fifty-pound, and seventy-five-pound masses. We set them off, measured car displacement, cleaned up, and set up for the next shot. When all was said and done, we determined that fifty pounds did not have quite the amount of energy needed to throw a car seventy-five feet. Yet seventy-five pounds was too much energy. In our slightly demented take of Goldilocks's analysis, the "just right" was somewhere between these two values. The official charge size estimate from South America authorities was sixty to one hundred pounds. Maybe there is something to this science stuff after all.

CHAPTER 5

Do You Want to See the Heads?

One thing you don't think about when joining law enforcement is the grisly turns the job can take. Shortly before Halloween in 2018 a concerned citizen brought what appeared to be a human head into a police station in Oakland, California. Police thought it a prank until closer inspection revealed "it was decomposed and had a little bit of flesh on it." Investigators interviewing residents of the apartment complex where the head was discovered were provided with insights such as "I don't know nothing about no head." One of the officers summed up the unique nature of the situation by noting, "I can say in my years of service I've never had a human skull delivered to the police station."

At least I start with a pretty fair idea of how a head came to wander off without a body in tow. In my line of work, when someone asks you if you want to see a tray of heads you know it is going to be a long day.

Every country eventually has its 9/11 moment. Indonesia referred to the attacks in Bali as their 9/11. Years later, Sri Lanka experienced such an event on Easter in 2019.

When Morocco's time arrived, the attack occurred in Casablanca on May 16, 2003. The Moroccan government, the city's emergency services, and the people were not prepared for the attack that came without warning on that fateful evening. Within a thirty-minute window starting at 10:00 PM, fourteen bombers waged attacks against five targets scattered throughout Casablanca.

The operatives were members of a North African terrorist group, Salafia Jihadia, with connections to al Qaeda. A dozen of the suicide bombers successfully detonated their devices, with two apprehended as they tried to deploy theirs. By the time the attacks were over, thirty-three people were dead and over one hundred injured. The attack represented the debut of suicide terrorism in Morocco and served as the deadliest attack in Morocco's history.

The targets hit represented an odd cross section of Morocco but focused mainly on Westernized and Jewish establishments. The deadliest attack occurred at the Casa de España restaurant, a Spanish-owned eatery, where bombers knifed a restaurant security guard and gained access to a packed outdoor courtyard. Once distributed throughout the patio tables the attackers detonated their devices and killed twenty customers. Two bombers attacked the five-star Hotel Farah. One successfully detonated his device, the second did not. Two bombers attacked an empty Jewish community center and two attempted to gain entry into a Jewish-owned Italian restaurant but ended up blowing themselves up outside the establishment. Finally, one lone bomber detonated his device next to a fountain 150 yards away from a Jewish cemetery.

Typically, a post-blast investigation focuses on figuring out what a bomb was made out of and how it functioned. In the case of Morocco, however, I spent a good deal of time making sense out of the sequence of events and target selection. Very little of it falls in line with the rigors of logic, but there are limits to what science can explain. Motivations of suicide bombers are often a mystery too deep to shed light upon with mere forensics.

No American citizens were killed or injured in the Casablanca attacks. But the Moroccan government called for assistance from the FBI. A single suicide bombing is a stress on any city's law enforcement agencies. Having a dozen such devices go off within thirty minutes creates utter chaos. The role of my team was, as it is so often, to serve as advisors. A senior examiner from the Explosive Unit and I deployed as a team. My job was to view the scene from the lens of science, his from the eyes of a forensic examiner and FBI agent. As we didn't have control of any of the scenes or a clear-cut mission (outside of "lend assistance any way possible"), we decided to take the thirty-thousand-foot view.

There were too many scenes to explore any individual one in depth. By the time we hit the ground, the local agencies had already "processed" the scenes. My partner, Leo West, and I met with the local FBI legal attaché (Legat) and decided we wanted to see all five scenes to get an overview of the situation. We could offer absolutely no insight without visiting the scenes firsthand. This was true triage forensics. There are times to dive into the weeds and times to head for the higher ground. We needed the view from the hill.

Processing a scene means different things to different folks. When the FBI processes a scene, we scrub it for every potential minute piece of evidence and do not release the scene until we are convinced we have exploited it to the maximum extent possible. This approach was not always followed internationally. Nowhere was this more evident than in the scene of the Casa de España restaurant. The passageway into the restaurant's interior was a narrow hallway, which opened onto the street where a guard was posted. We were greeted by a large smear of blood arcing down the doorway where the guard stood that night. It looked like a careless painter had taken a wide brush dripping with red paint and slashed a crescent from shoulder height to the floor. This was where the suicide bombers slashed the guard with a machete to gain entrance. After passing through the narrow hallway, the bombers emerged in the walled-off exterior courtyard. Four of them positioned themselves amongst the patrons enjoying a warm evening of food and bingo, and one signaled for all to set off their devices by yelling, "Allahu Akbar."

Fragments of tables were scattered across the floor in the courtyard. Scorch marks ran across the pavement and up walls. The chaos of terrified customers fleeing the scene, emergency responders attending to the dead and dying, and police processing the scene later left jumbled debris everywhere; there was little, if anything, I could do to make sense of the carnage. However, a cursory search of the patio turned up unexpected evidence still in place: fragments of twisted nails and ball bearings. There was no way that evidence like this should have been left at a post-blast scene. Every piece of bomb is picked up when we process a scene at the FBI. But this had not been our scene.

I pulled out a crime scene kit and collected some of the fragments, just in case. Nails as twisted as the ones I found would have been close to the main explosive charge. They would be a great source of potential traces of explosive. If the locals didn't want them, I wasn't going to let them go to waste.

I ambled around the courtyard and spotted something on the ground that did not look right. It was a small white fragment resembling bone. As I bent down for a closer look, I noticed it contained three teeth. I called Leo over.

"Am I really jetlagged, or is this a fragment of a jawbone?" I asked. Parts of a body are sometimes hard to identify out of context. It is easy to recognize teeth when looking at someone's smile. It takes the brain a little more time to process the same items as teeth when they are found lying on the ground disassociated from that smile.

My partner agreed with my assessment. I was not about to package up human remains the same way I did fragmentation. For one thing, these could have belonged to one of the victims, and as such needed to be returned to the family for proper burial, or they could have belonged to a bomber. At this point, I had no way of knowing. More importantly, there is no good way to try to explain to US Customs why you are returning to the States with a fragment of a mandible. We notified our local police escort of this discovery and headed off to the next scene.[13]

Part of the challenge of our site visits was that we had no one to tell us what had happened at each site. In some ways, this was good, as the interpretation of the local investigators could have been way off. It is hard not to develop some prejudice about events if someone gives you their theory. This is especially true if they present their theory as proven fact. On the other hand, having no earthly clue what transpired can really throw an analysis. The Hotel Farah site visit left me puzzled for months.

I hoped to identify each site's blast seats and measure damage to estimate the type and quantity of explosive that went off. At the Casa de España restaurant, blast sites were impossible to figure out. Farah presented a different kind of challenge. It had too many blast sites. From what I had been able to glean, a suicide

[13] As a side note, we would return to this courtyard two days later for reasons I cannot recall. The teeth and jawbone were still there.

bomber met resistance when trying to gain entry into the hotel. A wide doorway comprised of cinderblocks led into the hotel lobby. At one corner of the doorway the cinderblocks were obliterated. This area was closest to the explosive when it went off.

Surrounding the damaged doorway was an artifact called radial streaking. If you threw a water balloon at a wall it would splat water in a circular pattern, radiating outwards from where the balloon impacted. This is simple physics. When explosives go off, they send pressure waves outwards into the world. Some explosives produce copious amounts of superheated carbon atoms when they react. Basically, the blast produces ultra-hot soot traveling outwards from the center of the explosion. When this soot hits a flat surface, it creates streaks of carbon for the same reason the water balloon created streaks of moisture. Just like you can estimate where a water balloon, or paintball for that matter, impacted by tracing the radial streaks back to a center, the same is possible for blast. I could see from thirty yards away the streaks of soot emanating outwards from the portion of the hotel's doorway decimated by the blast.

To access the doorway to the lobby, with its heavy damage and streaks, you climb up about two tiers of shallow stairs from the turnaround drive in front of the hotel. A mystery greeted me at the base of these stairs. The pavement at the base of the stairs looked like someone had set off a bomb right at my feet. There were huge pockmarks on the pavement, and the lower marble steps next to the pockmarked pavement were shattered. This was nowhere near the obvious blast site at the doorway.

To create as much damage as I saw, the bomb had to be literally sitting on the ground in the hotel turnaround. But the obvious seat of the blast was a good thirty feet away and two tiers of stairs above me. The bombers were wearing backpacks. How in blazes did a bomb going off on the top of the stairs create damage so far away?

This made no sense. It wasn't until months later, when an agent sent me photographs from the scene that were previously kept from me, that the whole picture came into view.

There were, indeed, two bombers who went to attack the hotel. One bomber engaged in a wrestling match with one of the hotel security guards. During the scuffle, his bomb went off. This accounted for the damage I saw in the doorway. However, the second bomber was in close proximity to the tussle, and when his friend's bomb went off, it blew bomber number two backwards. During the explosion, the surviving bomber number two had his hand blown off and the backpack somehow dislodged.

The photo I saw shows the backpack lying on the floor of the hotel lobby. Coming out of the backpack is a length of black wire. This wire leads to a bloody stump of the second bomber's hand (which is still holding the pressure switch). Since nerve impulses do not transition well across an air gap, the surviving handless bomber was not able to set off his charge. He was taken into custody.

On the day of the explosions, the responding police had two fragmented bodies distributed across the scene and had to deal with a live bomb. Another photograph shows a dump truck parked in the paved turnaround in front of the hotel lobby with bomb techs sheltered across the street manipulating a rope line. This hook-and-line technique is referred to as rigging. The bomb techs had rigged a rope to the backpack and ran this line up over the bed of the dump truck. They were attempting to remotely pull the backpack out of the lobby, down the stairs, and up into the dump truck bed to take it out of the city and dispose of it in an area of their choosing. There was only one problem. The explosive in the backpack was hypersensitive, a fact the bomb techs were not aware of at the time.

As the bomb techs pulled the backpack, it dropped down about six steps. On each drop, it hit a tier of the obviously "rock-hard" marble. When it dropped down the last step and hit the street, the bomb exploded. Another photograph shows an explosion occurring at the foot of the stairs, oddly enough right in the vicinity of the mystery damage that baffled me for months.

Site three was more nebulous. On a narrow street bookended by the five-story Belgian consulate on one side and a small restaurant called Positano on the other, two of three more bombers set off devices. They did so as a security guard at the consulate saw them coming down the street and approached them out of suspicion. Debate ensued about whether they were aiming for the consulate or the Jewish-owned Positano restaurant.

We briefly visited the site, but aside from pockmarks on the street where the bombers met their end, there was not much else to see. Damage to the consulate could be seen up to the top floor. We were able to gain entry to talk to the diplomatic staff and take note of some of the interior building damage. We were given tours of the roof and interior rooms facing the blast. Sometimes you can get lucky and portions of the bomb blow up onto the roof of nearby buildings, or into windows shattered by the blast. No such luck was to be had.

Of all the targets, the Israelite Community Circle (a Jewish community center) was the most bizarre. This center was a gathering place for the local Jewish community, which comprised a very small minority in Morocco. The choice of this location as a target was puzzling, as the center was shut down for the Sabbath at the time of the attacks. The next evening it would have been packed with families enjoying food and festivities. No one could tell the full story of what happened there, as there were no good witnesses. We had to put together a picture based on physical damage alone.

The one thing that was obvious was that an explosion had occurred exterior to the building. The center was a two-story structure with a recessed doorway. All that was left of the doorway was shattered concrete blocks, which had held iron gates. Using the same clues that I described earlier from the Hotel Farah, the seat of the blast was obvious. Someone blew himself up in the locked recessed doorway. In essence, a bomber had given his life for the express purpose of gaining entry into the gated and locked center. Even with all the physical proof pointing to this undeniable conclusion, I still looked for evidence to dispute this. Why would someone kill himself to get into a building?

Inside the building was an equally bizarre spectacle. The interior of the Center had a great open area that was circled by a second-story balcony. Above the grand room hung a huge glass chandelier. I stepped over the mangled iron gates that were blown into the room by the first bomber—even more evidence that a device detonated in the doorway. The epicenter of another blast was in the center of the room. There on the floor, directly under the chandelier, was a circular scorch mark and radial pockmarks indicating the seat of a second blast. Fragment strikes marred the walls and a small stage just to the left of the doorway. The most unnerving part was the pattern of human remains.

Standing in the center of the room, I slowly turned in a circle. I could see blood splatters and pieces of flesh littering the walls all the way from the first floor up to the second-floor balcony. After watching his partner blow himself up to open the door, the second bomber stood in the center of an empty room and did the same. It was unusual to see a suicide attack with no victims anywhere to be seen. It's not that I can fully understand the mindset of someone who decides to martyr himself, but, standing in the center of that devastated room, I could not help but imagine the last seconds of someone who had witnessed the violence of a fellow human blowing himself apart, and then stood in the center of a deserted room in the dark to follow suit.

Bombers who wander into crowds with the intent to kill must feel something. Is it a sense of purpose as they close in on their victims, or a seething hatred for the people they target? Anger, hatred, and fear are primal emotions. They are within a scope of shared human experience that, although hard to rationalize, can be understood intellectually.

What does someone who is standing alone, with no victims in sight, shrouded in dark silence think before pushing the button that will end his existence? I cannot to this day conceive an answer. That question haunted me more than many others over my career.

The final attack site was put down as a logistical failure of the bomber. About a hundred yards away from a Jewish cemetery was a fountain used to clean oneself before making entry into the holy site. This was the last site we inspected. Basically, there was nothing to see. The fountain had a wall next to it, which was destroyed by the blast. One lone bomber had detonated his device there. The locals believed he had originally been supposed to meet up with some of the other bombers but had lost his way. Hearing the other blasts going off he sat down by the fountain and joined his compatriots.

His act blew the wall down on a nearby house where a small boy was sleeping. While we were there, the father came up to us to show us the cuts on his son's head. The boy was about the same age as my oldest son at the time. It is disturbing to see children injured by such acts of violence. When your brain associates these incidents with your personal life, it is harder to keep emotions in check. I have been blessed in that I have seen horrific carnage, but very little in terms of atrocities committed against children. I know law enforcement officers who take on that heavy burden. I am a scientist, they are saints.

Relationships move the world. This is the main reason the FBI puts agents all over the globe. Agents need to go out and meet their local partners and form bonds of trust. Nothing opens more doors and pays dividends that are as high. We caught a lucky break in the Casablanca case. Our Legat who oversaw relations with Morocco had just returned from a conference where she had met and formed a good relationship with the medical examiner from Casablanca.

After visiting the five physical bombing sites, I had gathered a fairly coherent picture of the types of devices deployed. However, no analysis is complete without a word with the dead (so to speak). So I was grateful for this connection.

Entering a morgue in the name of collecting evidence is an unpleasant task that etches indelibly on your subconscious—images that don't improve with age. In Morocco, estimating charge size across the various sites was fairly challenging. In an effort to gain some insight into size, we had to explore the damage done to the flesh and bones of the perpetrators.

To provide a simple instructional example of how analysis of a corpse can answer key questions in a bombing, I harken back to a case that I worked in Norman, Oklahoma. A young man had blown himself up on a park bench outside a packed college football stadium. At the time, speculation was rampant that the student in question was plotting to take his bomb into the stadium. This later proved false, but early on our only lead was a headless corpse and a gaggle of law enforcement officers who were fairly unnerved about responding to a scene where, despite the best efforts of those involved, they were faced with an unexplainable sequence of events. I wanted to estimate how big the explosive charge was that the young man deployed. I also wanted to parse together what he was doing at the time of the blast.

I had a few clues to go on before I visited the morgue in Oklahoma. First, a passing bus driver walking to his bus had seen the young man hunched over a backpack on a dimly lit park bench. As the driver got to his bus, about ninety feet away, an explosion erupted that almost knocked him on his face. He immediately called the police and identified the origin of the explosion.

When I got to the morgue, I was greeted by the body on the slab. The first obvious omission on the corpse was the head. That kind of thing jumps right out at you. The skin from his shoulders and chest looked like overalls someone had unbuttoned and let flop down over the stomach, exposing the muscle beneath. Also missing were his arms. To be more exact, the arms were not attached to the body. There were two arm segments on a separate table. These segments contained the forearms, elbow, and part of the upper arms. Missing was a head, two hands, and both shoulders.

Let's start at the top, literally. The head is an amazing product of evolution. Designed to protect the brain, a fairly valuable organ, it can take a fair amount of abuse. The largest section of skull present in this case was about the size of a small coaster, although its curvature would have made it impractical for application in such a purpose. The one thing I could be certain about was that the young man's head was in close proximity to the explosive when it went off. The missing body parts combined with the bus driver's eyewitness account painted the picture for me.

Put a backpack on your lap. Lift a bomb out of that backpack so the explosive is nowhere near your legs. Then raise the explosive charge up to your face and lean over it to better see in the dim light. Hunched over like this, the bomb will be touching your hands and be in very close proximity to your head at eye level and shoulders. When the bomb goes off, all of these things go rapidly away in the other direction. I may not know why the young man had the bomb positioned in this fashion, but physics pretty much dictated the physical location of both explosive and victim.

As I was working out the physics of the event, the lead medical examiner came up to me and said, “You've gotta see this.” If a medical examiner, someone who has seen just about every permutation of grotesque imaginable, feels something is noteworthy enough to elicit comment, and warrants showing off, you most likely don't want to see it. But I am a scientist and need to maintain a veneer of professional detachment.

On the side table sat three pieces of what I assumed were fragments that used to be attached to my victim. I stared at the three pieces for a few seconds trying to figure out what I was looking at. Then it hit me. I had patches of skin and hair sitting in front of me. With hair as a benchmark, I started to infer context. Just as I was about to make the association, the examiner could no longer contain his professional exuberance and he picked one piece up and proudly exclaimed, “It's his eyebrow!”

As soon as he put the piece back, the puzzle fell into place. He had arranged both eyebrows with a small fraction of the curly head hair from the deceased's forehead. The curly hair was only a small patch, as the young man had a receding hairline, and the tuft of hair was pulled off with a little of the forehead skin coming along for the ride.

Having put together all the pieces I needed, I could figure out both charge size and placement in Oklahoma. In Morocco, I had many more pieces but far less context.

The explosion sites in Morocco gave little insight into the size of the bombs. All of the sites had been cleaned up before we got there. Most occurred outside in open air. Basically, there was no "target material" from the scenes that could be used to assess pressures. The only reliable target materials I knew of were the bombers themselves. We needed to go to the morgue to view the bombers.

The morgue had three basic chambers. In the outermost room sat a small rolling dolly, much like the kind one would expect to find packed with books needing to be returned to the shelf in a library. A plastic sheet covered the top of the dolly and large blocks of ice weighed the sheet down. Vague outlines under the opaque sheet hinted that the dolly's covered contents were more macabre than a handful of Stephen King novels. With game face on, I pulled back the sheet to uncover a heap of hands and feet. The medical examiner beside me felt compelled to state that I was viewing the "hands and feet" collected from the bombing scenes. I don't know if he was unsure if I could identify body parts unmoored from their origins, or wanted to ensure me he wasn't in the habit of picking up random appendages and keeping them on ice in case he needed a spare part. Either way, unsure what the most socially adept response in this situation might be, I nodded and replied, "Yes, hands and feet."

I debated whether I needed to take an inventory of the number of limb truncations present. Then would I need to notate the number of left vs. right appendages tossed about together? I already knew how many bombers had blown themselves up, and there appeared to be more hands and feet than could be accounted for by the attackers. This meant that, while the exact accounting of dismembered body parts might make for some interesting cocktail party conversation, this number did not have forensic value. The medical examiner pointed in the direction of the bodies, through the doorway into the next room. And so, like in a haunted house, we went through to the next chamber of surprises.

In the next room, a wall was fitted with twelve roll-out cabinets stacked from floor to ceiling. Anyone who has watched any amount of crime-show drama on TV has seen such a wall. Typically, the medical examiner brings the police officers into the morgue and opens one of the freezer doors to roll out the body of one of the victims of a grisly murder. In those shows, the bodies always look fairly decent, albeit colored a bit blue to impart how cold they must be. But in Morocco, the bodies wheeled out from these drawers—what little was left of them—possessed a more crimson hue.

As the bombers wore backpacks, most of their torsos had been dispersed. Larger sections of the lower body will survive a backpack bomb, and there were plenty of bags of hips and legs in the morgue. It is hard to get a good estimate of charge size based only on the damage to a body, but a decent ballpark estimate can be achieved if you have enough bodies, or fractions thereof. The damage I saw spoke to ten- to twenty-pound charges easily. As I went through the third body bag, reminiscing fondly of the far-less-gruesome cart of hands and feet, the medical examiner came up to our team and asked, "Do you want to see the heads?"

In my line of work, I sometimes step back from myself internally and observe the situation as a neutral third-party observer.

"Did he just ask if I wanted to see 'the heads'? I had to mishear that. Perhaps he wanted to know if I needed to use the head? No, that makes even less sense. He seems to be staring at me. I wonder why? Hey, idiot, he is waiting for a response. Stall."

I looked up from the body bag that looked like extra set dressing for the movie *Saw*. He asked again, this time adding, "Do you want to see the heads of the suicide bombers?" With a nod of my head, we went into the next chamber to continue the grisly tour.

This room was an autopsy exam area. A standard metal table stood in the middle with a large light looming above it. To my left in a corner, a large metal tray had twelve heads neatly arranged upon it. Lined up in two rows of six, each had a number written on a piece of tape stuck to the owner's forehead.

For the most part, the heads looked pretty good. This, of course, is a relative statement. To be more exact, they looked good for heads that had been ripped off a torso by an explosive blast, launched into the air, and crashed down on the hard ground. One of the heads was still attached by a string of skin to a portion of a shoulder and an arm. This head had been arranged so that it looked like the bomber had decided to rest his head in the fold of his arm to take a nap. The illusion made for the creepy icing on top of the horror-movie cake being served up on the platter.

Damage to the heads was consistent with the charge size I had gleaned from looking at the remnants of bodies on the slabs in the previous chamber. I leaned over to my partner and pointed out to him that one of the heads had a goatee. While the presence of a goatee was of no value to a scientific analysis—the goatee won't help hold the head on longer or provide aerodynamic resistance and slow the head down in its parabolic transition away from its old host towards the awaiting embrace of mother earth—it would, however, serve as an investigative lead. Suicide bombers of certain faiths undergo ritual cleansing before they kill themselves. Part of the ritual is to make sure you have a clean shave. The presence of the goatee stood out.

After reviewing all the separate collections of anatomy, I was able to make a couple assessments about the bomb. Placement was simple as it corresponded to the point of most physical damage. Backpacks made perfect sense based on this. Size estimates from body damage are a little more challenging. However, heads tend to remain attached in a certain charge-size bracket and detach within another. An absolute upper limit can be estimated based on what a normal person could carry without eliciting attention stumbling down the street. All things considered, the bombs seemed to fall within the twenty-pound range.

By our third day in Morocco, the FBI was trying to figure out what we had left to do to lend assistance. We had seen the sites of all attacks, had examined the bodies of the bombers, and had formulated about as good a picture as we could expect to develop. The last thing to do was visit the forensic lab. We were on the way to do that when we got diverted on a side mission.

Sometime earlier that morning, or the night before, the local police had located the bomb factory where the Moroccan bombs had been made. They had collected all the chemicals and explosive production equipment and were carting it off to a local police station. As the lead chemist for the group, this was a situation in my wheelhouse. We diverted directly to the police compound. What we found there was a flurry of activity swerving down a bad road.

The chemicals the police had recovered at the bomb factory were those needed to make TATP (Triacetone Triperoxide), which is highly volatile. Though it's come into public consciousness through relatively recent ISIS attacks in Paris, Brussels, and Manchester, the use of TATP by terrorists goes back much further.

TATP was discovered in 1895 by accident by the German chemist Richard Wolffenstein. Wolffenstein did not want to make an explosive. He was actually trying to chemically react a component of hemlock to create another compound. In this process, he dissolved hemlock leaves in Acetone to pull out the other compound and added Hydrogen Peroxide to this mixture to react with the extracted material. He let this solution sit for many days. Although the Peroxide never reacted with the hemlock constituents, it did find a companion in the Acetone solvent. Eventually, a white solid came out of the solution, which, much to Wolffenstein's dismay, was a hypersensitive explosive.

Most serendipitous discoveries of explosives led to their incorporation into military or commercial applications. This was not the case for TATP. This material was so sensitive to external stimuli, and so chemically reactive, that no one wanted anything to do with it. It quickly vanished from the chemical landscape, not to emerge for another one hundred years.

When al Qaeda awoke TATP from its historical hibernation, they did so for reasons of practicality. Al Qaeda limited their usage of TATP to small charges, typically less than a portion of an ounce, to set off larger charges. Put TATP in a small paper tube and you re-create the detonator used in Richard Reid's shoe bomb. It was in this fashion that TATP entered the terrorist explosive landscape. But there was another evolution between Reid's shoe and the attack in Casablanca.

In the late 1990s, the Palestinians took the work of al Qaeda one step farther.

Whereas al Qaeda bomb makers only made minute quantities of TATP for use in setting off larger quantities of much more stable explosives, the Palestinians decided to cut out the middleman and ramp up TATP production to levels that are purely terrifying.

The conflict between Palestinian terrorist groups and Israeli forces is long and complicated. Through the course of this conflict, Israel placed many restrictions on the types of chemicals allowed to be brought into the Palestinian territories. The TATP advantage, then, is that it is created from combining three very basic and common chemicals: Acetone and Hydrogen Peroxide (in fact, TATP[14] is sometimes just referred to as Acetone Perox-

[14] I typically don't discuss explosive recipes but this one is so common these days anyone can find the ingredients with no challenge. For those thinking of giving the preparation of TATP a shot, most of the time I run across people making it is when they show up to the hospital with fewer body parts than they started off the day in possession of.

ide), and acid. Acid, referred to as a catalyst, is used to speed up the reaction. In chemistry, a catalyst is a species that encourages a reaction, but does not get changed by it. Typically, the acid used in the prep of TATP is Sulfuric Acid (or battery acid). Acetone, Peroxide, and battery acid are three very common chemicals, and they were chemicals the Palestinians could easily access.

The Palestinians had limited access to chemicals they could use to make a main charge for their bombs, but they could get huge quantities of the precursors needed to make TATP. For this reason, some intrepid terrorist groups decided to just make huge batches of TATP as the main charge. The Palestinians began to produce hundreds of pounds of TATP. It was used to make a wide variety of IEDs. In one case, a watermelon was hollowed out and stuffed with TATP to produce a bomb found on a bus.

There was one big disadvantage to scaling up TATP for use by the pound. TATP is sensitive enough that routine handling can sometimes add enough energy to piss it off. Just like Bruce Banner, you won't like TATP when it gets angry. Many Palestinian bomb factories blew up in the process of preparing this material. It may seem hard to understand how any rational human would continue to work with such a highly vindictive material, but determined bomb makers are willing to take risks. The more determined they are, the bigger the risks they partake in.

I was told by Israeli bomb techs that in some cases they would respond to a Palestinian bomb factory to find dead, and naked, bomb makers. It took them some time to work out the naturalist aspect of the scene. It is one thing to stumble across nudists playing volleyball, it is quite another to find them in a bomb factory. As it turns out, the Palestinians had developed such a fear of TATP that some had completely disrobed when working with it out of fear that the static charge on their clothes would set it off.

During the early 2000s, the Palestinians were the only group producing mass quantities of TATP for their main charges.

Eventually, they tired of the toll this material extracted on their numbers and went on to other explosives. In Morocco, TATP made its first leap to the wider world stage.

The Moroccan police in their attempt to make the residual TATP safer had created miniature bombs. Cases of empty chemical bottles sat in the police parking lot. Some bottles were clearly labeled Acetone, while others had *eau oxygénée* printed on their sides. I might not have understood French, but I was pretty sure *oxygénée* was synonymous with oxygen. Also, the bottles were clearly labeled 12 Volume. Only one chemical lists its concentration in this weird archaic fashion, and that is Hydrogen Peroxide used for bleaching hair. There was no doubt that this was a large-scale TATP lab. It was also the first large-scale TATP lab to be found outside the Palestinian territories.

Coincidentally at the time, I just happened to be the leading US authority on TATP, and what I was seeing in the parking lot was tripping every Spidey-Sense I had developed regarding the material.

The police had taken every bottle out of its case and were filling them with water. The bottles were all coated on their insides with large crystals of TATP. Out of concern that this coating could spontaneously detonate in the parking lot due to UV radiation from sun exposure, a French forensic team who was assisting the Moroccans recommended adding the water in an attempt to stop this from happening.

Sometimes, I have found, you have to apply finesse in situations when you want to strangle someone. I almost blurted out, "If you left the damn bottles in the cardboard cases they were all packed in, they would not be in the sun in the first place." While such an outburst might have been cathartic to me, it would not have been productive. However, I knew TATP was not water-soluble. I also noted that the addition of water was actually creating a side effect not anticipated by the French team.

The bombers had filled the bottles with the reactive solutions to let TATP come out over time. They then filtered off the TATP from these bottles, which resulted in a coating of TATP in each bottle. By adding water, the TATP, which had been spread thinly along the sides, washed off and floated up to the neck of every bottle to form a solid plug. In essence, by filling the bottles with water, the bomb techs had created mini charges of TATP in the neck of every one.

I pointed out to the group that they had just created dozens and dozens of explosive charges scattered across their parking lot. That point did seem to get their attention. The French team seemed a little taken aback by my technical assessment, while the locals stood there eyeing the sea of misfortune they had created. Finally, the commander of the bomb squad asked what I advised they do. It was just one of many impossible situations I seemed to get dropped into. Frankly, the damage had been done. There was no guaranteed safe way to back out of the situation. I told them to get everyone not needed out of the lot, as a crowd of observers had gathered to witness the festivities. I told them to stop adding water and creating more bombs. Finally, I advised that they get on a bomb suit—they had to treat each bottle like it could explode at any moment—and pack each bottle up, take them to the desert, and blow them up.

My TATP expertise suddenly made me both needed and wanted. Morocco had never seen TATP, and I had spent more years studying the material than just about any other scientist in the world. It was time to put on my teacher's hat and ring the bell to bring class into session.

Up to this point our team had been at a disadvantage. We did not speak the language, arrived at each scene well after it had been processed, and didn't seem to be valued as a resource by the Moroccan police. When we arrived in the lab, that all took a 180-degree turn. We were greeted by the head of the lab and shuffled off to a conference room to have a private conversation. The presence of TATP had taken the local lab totally by surprise. They had no idea what they were dealing with. Our team had practical experience with it and years of research into its properties. All of a sudden, people wanted to talk to us.

The head of the local lab brought a couple of his chemists into the conference room to have a side discussion. I began to explain some of the properties of TATP when I was cut off in mid-sentence by one of the scientists. I still recall quite vividly him putting up a hand to stop me and asking, "Is this information you have just read in a book, or do you have real experience with this explosive?" Had he asked me this question in a polite manner, and had I not just gone through multiple days of being politely held at bay, I might have been inclined to answer in a fashion more driven by diplomacy and tact. However, by this point I had had enough. They were going to get someone killed because they did not understand the threat they were dealing with, and I had no patience to deal with someone trying to erode my credibility out of ego. I addressed the lab director and his scientists very abruptly and noted that I was the first US scientist to make a study of TATP. I had done all the US research on its safety and its detonation characteristics. I had produced all the training material US bomb squads had at the time in how to deal with the material. I ended by noting I would be happy to pass that information along if they were interested, or I would be equally happy to go back to the United States and let them figure out how not to blow themselves up without my assistance. There were no more questions aimed at my expertise. Sometimes, you have to be an ass.

I taught the local lab scientists about TATP for several hours. TATP can be tricky to analyze post-blast and the FBI had developed procedures to do just that, so communication with the Bureau was crucial. Our knowledge in this area spawned a new spirit of cooperation with our counterparts. I worked with the bomb techs to give them all the insight I could into how to deal with the material, and ways that the bad guys could weaponize it. Based on the number of bottles of precursors present, I was able to give an estimate into how much TATP the lab could have made. In the end, it appeared the bombers had enough precursors to produce about two hundred pounds of the volatile material.

I was amazed by both the ingenuity of the bombers and the fact that they got through their prep without eradicating themselves. They had rigged large gallon glass jugs with an apparatus made from a sewing machine motor attached to a bamboo stick. This device allowed them to stir the chemical mixture while ingredients were being combined. It also allowed them to control the speed of agitation along the way. They had added a couple of other ingredients to the TATP in an attempt to make it more amenable to handling. The exact ingredients are beyond the scope of a tale like this, but they mirrored some of the things that I had seen the Palestinians do years earlier. What was most impressive was the way the terrorist cell adopted TATP in such large quantities seemingly out of the blue. Up to that point, we had seen other terrorist groups adopt small amounts of TATP and then slowly get more proficient at its production until larger batches were produced. The Moroccan bomb makers went from zero to sixty without starting at zero. I am still unsure where they got the knowledge for the operation, but they seemed very well informed. Bad guys talk to each other, and the internet allowed a closer alliance of those wishing to engage in explosive production and bomb making.

I have been to many countries. During my travels I have not always seen these nations in their finest moments or interacted with their people in the most relaxed hour. Most of the locations I have been deployed hold no attraction for return. Morocco is one of a small handful of exceptions to that statement. There was something about Morocco that I found myself drawn to.

After all the drama of chaotic crime scenes, the carts, drawers, and trays of dissociated body parts, capped off with my chance to actually do some good, we spent a final night in a Moroccan bar to unwind.

The room was dimly lit. Windows were open to let in a breeze from across the water. Looking out the window I could see the lights of the Hassan II Mosque. With golds, white, and turquoise glowing across the inlet the mosque radiated a calm. In the background the strains of Moroccan folk music flowed. With Spain across the water and the African continent at my back I could feel the wash of both cultures like waves hitting the shore. It was serene, and mysterious, and for a moment perfect. I hope one day to return and maybe slip back into that feeling once again.

When the Moroccan attack occurred, I was still fairly new in my career at the FBI. My boss told me that FBI director Mueller had taken a personal interest in the bombings and wanted a briefing. Normally, Leo, as the senior agent, would give the briefing, but since the incident involved TATP, and I was the resident US expert in the material, it was decided that Leo and I would go together to our headquarters to brief Director Mueller. It was my first time briefing an FBI director, but it would not be my last.

It is funny the things that you pick up from your parents without even thinking about it. My father was an English teacher in the mid ’60s and, being the newer generation, refused to wear a tie. To make matters worse he had my mother custom sew him “suits” made from an odd assortment of colors and paisley designs. My dad didn't care. He felt that if the existing establishment didn't like his outfits, they could fire him. Without realizing it, I internalized part of that message.

Today the only time I wear a tie is when I want to deliberately blend in somewhere. However, during my years in New Mexico I started the habit of wearing bolo ties. I decided that I would meet Director Mueller as myself. I did so in a maroon dress shirt and a bolo tie adorned with a rare piece of "white" turquoise.

I did not realize at the time that Director Mueller absolutely hated to see his executives in anything but white dress shirts. Even blue dress shirts would rub him the wrong way. Leo turned to me after our briefing and said, "I can't believe you just briefed the FBI director in a bolo."

At this point I should say that our briefing went very well, and Mueller was very complimentary about the work we did. In fact, a year later he would be kind enough to award me with one of the FBI's highest honors. That is another story. Being a scientist gives me some leeway to be myself, bolos and all. Mueller respected the insight I brought to the table and forgave my choice in neckwear.

Years later an FBI Lab director decided I could not participate in a briefing with Mueller because I did not have a suit coat with me. I never told him the story of my first briefing with the director. In life there is form and there is substance. Those who obsess on form oftentimes get caught in the shallow end of the pool.

CHAPTER 6

The Not So Magical Mystery Tours

In an era where whatever you want to know is just a "hey, Siri" away, it's easy to feel informed. And it's easy to form opinions, secure in our "knowledge." But my work in the dark underbelly of the world has brought into stark relief that, often, what we think we know are just words on a screen. There are many things you find you never truly knew, once you actually come to understand them.

I should have known that this trip was going south from the moment my gear vanished, which, in my line of work, is no small inconvenience. Selecting gear to take on a deployment is tricky and time consuming. In many countries, I can depend on local police to have evidence bags and gloves, as well as a variety of rather mundane crime scene processing essentials. But in January 2004 I was heading to Bangladesh to provide technical guidance to the locals who had just rounded up some suspected bomb makers. Knowing that the financial hardship of the region would mean a lack of supplies, packing light was not an option.

So, I packed up the largest FBI Pelican case I could carry (or, more specifically, roll) with everything I could think of needing to process a bombing scene.

My kit contained swabbing supplies, evidence-packaging supplies, a wide assortment of measuring devices, a camera, headlamps, flashlights, pocketknives, lighters, spatulas, tongue depressors, box cutters, multitools of various sizes, magnifying glasses of a wide assortment of powers, work gloves, flight suits/ overalls, and a great number of other small and assorted tools to fill any space left. Being a bomb examiner is kind of like being a car mechanic or plumber. It is a dirty job that requires a wide range of tools, and you never know which ones you'll need until you start the job.

After twenty-four hours of travel time to cross eight thousand miles, I stood with growing unease as other passengers collected all their belongings and drifted away into the steamy night while I continued to wait. Then the luggage belt came to a stop and left me with a noticeable lack of a crime scene kit.

Instead of heading to a hotel I had to head over to the airline lost baggage desk with our embassy fixer and attempt to explain through a series of translations and gestures what I had lost, and its importance to our mission. An hour later, convinced that I would never see this case or its contents again, I departed wearily for the hotel.

Arriving in a foreign country can be disorienting. After a day of travel with little sleep and a middle-of-the-night arrival in a torrential downpour, I was extremely disoriented. I remember a dark swirl of tropical trees, small trucks, rickshaws, and, eventually, the iron gates of our hotel, protected by armed guards. I realized there would be no morning walks around the city.

It was about three-thirty in the morning by the time we hit the front lobby of the hotel. Morning meetings would start at about eight o'clock. We planned to all meet for breakfast at about seven.

It may seem odd that I recall these times years later, but, in most of my travels, I hit ground about seven to nine in the morning and go right to work. I now had three hours of potential sleep before the start of a new day. That struck me as a welcome luxury.

I went up to my room, still cursing the loss of my gear, but somewhat comforted by the thought of at least a few hours' sleep. The rain was sheeting down the bedroom window as I tossed my bag into the corner and partially undressed to crawl into bed. Occasional peals of thunder echoed outside. From my second-floor window, I could barely make out the metal gates through the driving storm.

Just as I put my phone down on the dresser and turned around to walk over to the bed, a brilliant, blinding flash lit up the entire room. One second later, the loudest explosion I had ever heard (and that's saying something) cracked outside, shaking the entire room. I was immediately plunged into pitch-black silence. No power.

My first thought was, "That had to be a fucking car bomb." Then, as my eyes adjusted and I had a moment to process, the scientist in me looked around, noticed the still-intact windows, and knew there was no bomb. Not to be dissuaded, the adrenaline-soaked portion of my brain functioning on no sleep kept insisting it was under attack. I soon regained enough sense to pull out the flashlight[15] from my pants pocket and find my cell phone.

I called down to my partner, Rich Stryker, asking if he was OK. He was. A call to the front desk revealed that lightning had struck a transformer just outside the hotel and it exploded. The power would return but flickered unsettlingly for the rest of what remained of the night.

---

[15] In this business, it's like the *X-Files*: 1.) Always have a flashlight handy. 2.) Trust no one.

Later the next day our team went to what was described as a "police station." I found myself in a barbed-wire-wrapped compound situated within an already high-security compound, like a locked box inside a fireproof safe. We had some cursory conversation with a mid-level ranking official as the power was occasionally flickering around us. My skin crawled when the official congenially turned to offer Rich and me a tour.

Our host took us down a series of narrow, dimly lit stairwells to the basement floor, where he pushed open a stout door that led straight into a hallway from a Stephen King novel. I had zero desire to take another step, frankly. But I couldn't just abandon the mission, despite being hazy on what that mission really was. It involved a bombing...or a plot of a bombing; we'd yet to have all the details filled in.

We followed as a guard led us toward several smaller rooms off a main corridor. In the first room, a chair sat next to a simple table that had been pushed into the corner and held what looked like a very heavy motorcycle helmet. I glanced up to note a large metal hook hanging from the ceiling. I decided I would not ask about the hook. I didn't really want to know.

The next room was bigger—about ten feet by twenty feet, the size of a generous living room, but not what I'd call inviting. There were more hooks, this time on the walls and on the ceiling, and not the kind that beckon you to hang up your scarf and hat. Large metal bars and rings about an inch thick evoked the rooms in *The Texas Chain Saw Massacre* rather than Ward Cleaver's home.

In the far corner was what looked like a dinner table crudely put together from rough wooden slats and iron bands. The table was about six feet long and three feet wide. At the corners were welded shackles. It took me a second to process what I was seeing: I had come face to face with a homemade rack. A large iron wheel was connected to the table through a series of gears.

The tour guide enthusiastically described how the table was used. Wrists and ankles could be shackled to the corners and then the wheel could be used to spin the table around so that the interrogators could "beat the feet." I am rarely a man who finds himself devoid of words. Typically, I can find something to say relevant to a social situation. However, try as I might, I would find no suitable comment to make to our host. The voice in my head kept repeating, "Nobody expects the Spanish Inquisition." Because, honestly, what the hell else does your brain do but try to ease the horrific imagery? But outwardly I kept silent.

To break the silence, my team member Rich, after giving me a "well, this is fucked up" stare, turned to our proud host and asked, sarcastically, "Which one of these contraptions do you find works best for you?"

A police officer approached asking if we wanted to see the suspect. I, for one, did not. Because this really wasn't my job. Foreign counterparts see all FBI folks as the same. We are all FBI agents to them. I am a scientist. Agents do interview and interrogation. I do not. In the most relaxed of situations, I am not about to sit down and start asking a suspect in another country questions. I sure as hell was not going to do it with a chamber of horrors as a backdrop.

Before I could come up with some way of addressing my concerns, our host brought in the detained suspect, believed to be one of the henchmen in a bomb-making factory. I couldn't take my eyes off of him. He was small and emaciated, dressed in filthy, torn clothing. He had been shot in the leg, though nobody could tell us exactly by whom or why. And he was hooded like a falcon.

So, there I was, sitting in a torture chamber with local police and a man in handcuffs and leg shackles with a bullet wound in his leg and a black bag over his head. Perhaps reading my abject revulsion of the situation as fear, the police officer pointed to our suspect and said, "Don't worry. He and I have become very good friends over the last couple of days." He then pushed him down in a chair and, raising his voice to the suspect, asked, "Haven't we become friends?" He then removed the hood to allow a more intimate discourse.

My blood ran cold, as I wondered if I was going to have a front-row seat for a matinee showing of *Midnight Express*. Luckily, a senior FBI agent came into the room and took over. The polygraphers were ready to interview the "miscreant," to use the terminology of our hosts. So the police hooded him back up and took him to meet our teammates. The senior FBI agent, God bless him, engaged the local police in a conversation on how the FBI could send a team of folks over to teach some other, maybe less barbaric and inhuman (though he diplomatically refrained from those phrases), methods of interview and interrogation that we had developed over the years.

The tour over, and the suspect whisked off, we bomb guys had an excuse to get out of this David Lynch screenplay. Rich and I headed back to the hotel. We were on our first beer when the polygraphers joined us and shared unsolicited information about the torture chamber we had just witnessed.

The full head helmet on the table was equipped with intense lights so the suspect could be fitted with a portable version of the bright desk light cops use in dime-store detective novels. The hooks were to hang people up by shackles to make them more compliant. The wheel on the rack not only moves the platform from a vertical to horizontal position to allow free access to the soles of feet, but also spins the rack in circles at a high rate of speed, like some demented amusement park ride. Additionally, an electric shock device could be fitted to suspects that, when the electric current was administered, drew on the building's current, causing the building lights to flicker.

Shortly after our team left Bangladesh, its government established the Rapid Action Battalion (commonly known as RAB),

Bangladesh's elite anti-crime and anti-terrorism force. RAB's torture methods included everything I had been told about and then some, including boring holes with electric drills. Their working motto was, "Criminals cannot have any human rights."

Dehumanizing an adversary is the first stage in abdicating one's own humanity. Having seen how little human life means in some places across the world, I have come to cherish it more. The same can be said about human rights. When I heard that the US had engaged in activities that treaded these same dark waters, it was one of the sadder days in my career as a public servant. Bangladesh cemented that lesson firmly in my conscience. I finally understood what these concepts meant.

It is one thing to engage in abstract debates about torture, or tie a ribbon on the pig and call it enhanced interrogation; it is quite another to be brought face to face with it. You don't know until you really understand. That's a life lesson we should all carry. The more important the issue, the more essential it is that you really understand what you already "know."

CHAPTER 7

When We Become the Bullseye

My father (and quite a few others) often asks, "Aren't you afraid the bad guys you're identifying will come after you?" The short answer is no. Of course, those of us in my line of work all have a healthy amount of fear in our daily jobs. I have had three friends in the profession mangled by explosives, losing fingers and sustaining permanent injuries such as hearing loss, and one close colleague lost his life. But we tend to be more wary of the explosive nature of the materials themselves rather than those who have them in their possession at any given time. Also, bombers—if they live—don't generally care if you know they did it. Terrorist groups clamor to take "credit" for acts they didn't even commit, and lone bombers want everyone to know their grievances. After all, their main mission is to make the mighty Goliaths know that the Davids of this world can mess them up. By and large, they're not after us.

That's not to say that we're never directly in harm's way. By nature, bombers are dangerous, and sometimes deranged, individuals. As scientists in the field, we have little protection against them in real time. They also set off an angry chain reaction in the community under attack, where those coming in to help are sometimes perceived as the enemy.

The one time I was sincerely afraid for my life was in Beirut in 2005 following the bombing assassination of anti-Syrian journalist Samir Kassir, where I led a team for evidence recovery. Multiple FBI teams would go into Lebanon that year. Mine was the first team invited into Beirut after the Syrians "departed." My friend and colleague, Michael Leone, and his crew came in after the next attack. For the next five months, our unit was on near constant rotation in and out of the country, as attacks kept occurring. At some point, it dawned on us that we couldn't keep showing up every time something bad happened. People were starting to realize who we were. Intel reports began to indicate that we were becoming the target.

On June 2, 2005, journalist Samir Kassir got into his Alfa Romeo, which was parked in a Christian part of Beirut. A bomb placed under the driver's seat of his vehicle exploded as he started the engine, killing him instantly and wounding a woman who was accompanying him. At the time of his death, Kassir was forty-five years old and one of the most prominent liberal Lebanese journalists.

Ordinarily, the bombing of a single individual who was not a US citizen would not result in a call for FBI assistance. However, Kassir was different. The political vortex that existed in Lebanon at the time was complex, fraught with tension, and just chaotic enough to suck an unwitting bystander like myself into its maw.

The chaos started that winter. A few months before Kassir's assassination, on February 14, 2005, Rafic Hariri, the former prime minister of Lebanon, was assassinated by a massive roadside truck bomb estimated by some to have contained two tons of high explosives. The bomb went off as Hariri's convoy of six vehicles passed the St. Georges hotel located near the Lebanese capital's waterfront. The blast annihilated the convoy of vehicles carrying Hariri, killing him, eight members of his entourage, and thirteen bystanders. The explosion left a thirty-foot-wide crater in the street, destroyed twenty cars in its vicinity, and injured one hundred additional people.

Prior to the attack, Hariri had stepped down as prime minister amidst disputes with President Assad of Syria over Syrian control of Lebanese politics. Hariri was focusing on upcoming elections, which had the potential to vote in a new government and bring him back into a position of power. The magnitude and complexity of Hariri's assassination led people to suspect that the attack was orchestrated by Assad and his regime.

Due to the tremendously high profile of this attack, its investigation went well beyond the level of the FBI providing some assistance. The United Nations itself established the Special Tribunal for Lebanon in The Hague to pursue the investigation. The UN sent a team of experts to process the scene. Forensic analysts from the Netherlands, some of whom I worked with over the years, reassembled what remained of the bomber's flatbed truck. These experts also collected hundreds of body parts at the site (a task I do not envy). Most were identified as belonging to Hariri, members of his entourage, and other known victims. However, about one hundred genetically compatible samples could not be linked with the identified victims.

From experience, I know that when unidentified body parts are recovered from a scene the most logical conclusion is it's a potential suicide bomber. In this case investigators had other clues that the parts were from someone very close to the bomb, and not some as-of-yet unidentified victim. In particular, the scatter pattern of those parts showed that they belonged to the person closest to the explosion. In addition, the small size of the parts also showed close proximity to the blast. The largest body part found was a nose. Other investigators tried to use the shape of the nose to infer the bomber's potential nationality.

Hariri's murder had a massive ripple effect. In the following months, massive protests against the Syrian government's twenty-nine-year occupation of Lebanon erupted in an event that become known as the Cedar Revolution. A cascade of events followed, including the resignation of the pro-Syrian government in Lebanon on February 28 and the initiation of withdrawal of Syrian troops on March 8. The popular demonstrations reached their apex on March 14 when the Lebanese people held the largest anti-Syrian occupation demonstration in Lebanon's history with over a million demonstrators in attendance. By the end of April, the Syrian government pulled out all of their troops and dismantled all of its intelligence stations in Beirut and north Lebanon.

Rapid change does not come easily. While the Cedar Revolution liberated Lebanon of Syrian control, it also divided Lebanon's political parties into pro- and anti-Syrian factions that endure to this day. Samir Kassir became a prominent figure in this revolution. As a voice for the anti-Syrian sentiment and a respected and influential columnist at the leading Lebanese daily, he was highly visible across Lebanon. Kassir had boldly blamed Syria for the assassination of Hariri, called for the resignations of pro-Syrian politicians, and served as a founding member of the Democratic Left movement. Kassir was viewed as a galvanizing force behind the transformational mass demonstrations that occurred after Hariri's death.

When Kassir himself was killed, his visibility as an anti-Syrian activist made it easy to blame the attack on pro-Syrian forces. His death was used as evidence of continued interference of Syria in Lebanese politics and led to further calls for the removal of the pro-Syrian president. The Lebanese government was desperate to show its people it was serious about investigating Kassir's assassination. It wanted to prove it was being open and fully transparent. So it reached out to two neutral countries for assistance: the United States and France. The French sent their top team of investigators from their national lab. For some reason, the FBI sent me.

I led the first FBI team to enter Lebanon since bombing attacks against our military installations and embassy decades earlier. At the time, I was fairly unversed regarding the political situation. Perhaps my ignorance about the hornets' nest I was being tossed into served me well. Dammit, Jim, I'm a doctor, not a diplomatic envoy. But by the end of the deployment, I would be a little of both.

The bomb underneath Kassir's car detonated at 11:45 AM local time. This translated into 2:45 AM Quantico time. (Quantico is home to the FBI training academy and laboratory.) Typically, when a call for assistance comes from another country a long line of approvals has to happen, and days can pass before the FBI hits the ground. My deployment to Lebanon was the quickest turnaround in my career. When I got into work that morning, my boss instructed me to turn around and pack a bag for Beirut.

I had pre-staged bags of gear ready at all times. Deployment gear is easy to have ready to go in a travel case. Regular clothing was also easy to pre-stage. For the most part, I was always ready to go to a hot climate. It seemed everywhere we got deployed was hot.

I had tickets waiting for me at the airport when I got there that same afternoon. My previous experience in Bangladesh with "misplaced" gear made me paranoid. Not only did I have my standard full-size kit, I had developed an emergency back-up deployment kit that I kept in a large backpack I carried with me. It carried the same items as were in my large case, but in "fun-size" proportions.

Going through security with measuring tools meant for a bomb scene investigation can make airport travel both interesting and tricky. I have learned that there is never any guarantee how many times you may go through security when traveling overseas. In the US, a traveler enters the system and is good all the way through any connections. In Europe, there can be a security point between terminals, and then an additional X-ray system at the departure gate on top of that. Some countries even set up a random baggage check at the gate to ensure security. Travel to Israel, and a nice conversation with a curious security agent will be added to all of that.

My plane routed through France. There, one of the multiple layers of security took notice of one of my measuring tools, called a caliper. Its ominous shape—that of a metal ruler with two small flat jaws—struck terror into their hearts, and even a simple explanation and demonstration of its use did nothing to allay the terror it induced. I told them I wished to talk to the airport police.

Most terrorists do not ask to be brought to the airport police, so I was viewed as more of a curiosity than a threat. I showed the police my US diplomatic passport and FBI credentials and proceeded to give a lesson in precision measurements. I could tell they were still skeptical. So, I suggested that the police keep my tool and accompany me on the plane and hand it to the pilots for safekeeping. This way, everyone could be assured I would not get up midflight to try to take over the aircraft through strategic application of precision measurements. This seemed to make everyone happy, and I was allowed to board with my escort at the very last minute.

Ironically, upon departure from the plane, I was so tired from the multiple flights that I forgot to retrieve my tool from the pilots until I got to the top of the flight ramp. This meant I had to engage a gate agent in Beirut and try to explain why I needed them to go talk to the pilot for me. But by this point, I was sooner going to leave behind a kidney than that damn caliper.

Getting into a country on deployment is a leap of faith. Many countries require a special visa on a passport for entry. Most of the time, I was sent off with mere hours' notice with the assurance that "it will be taken care of on the other end." I came to recognize that as code for "with any luck someone will let you through customs." When I got off the plane in Lebanon, I was luckily met by some local authorities whose job it was to escort me through their system. But the look on the customs agent's face as he thumbed through my passport washed away any relief I had. As his lazy gaze shifted from page to page, he suddenly stiffened, staring at something in disgust. He had apparently uncovered a mortal sin.

The customs agent was staring at my old visa from Israel, a past deployment. The Israeli visa sticker—a large menorah—takes up an entire passport page. Certain countries do not recognize Israel. Oh, they can identify it on a map if pressed, but officially admitting to its existence as a sovereign nation is a step too far. Anyone sporting evidence on their passport that they have visited such a non-existent nation threatens the cherished tradition of denying said nation. Had my passport had visas from Narnia, Pandora, or Atlantis I would have been fine. Mine, however, was marred by the visa from Never Never Land. Now I was faced with an insulted bureaucrat who had been trained to view anyone with such documentation as (at best) a spy.

My gate escort immediately engaged in an animated exchange with the offended agent. The escort turned to me and confirmed the issue, asking, "You have been to Israel?" My response was a tired and irritated, "Multiple times." The escort noted that they typically don't allow people into their country with Israeli visas. At this point, I lost some of my tact.

"I can respect that," I said, "but the president of his country had requested FBI assistance. I am more than happy to either provide the requested assistance or get back on the next plane and head home."

Quickly, the escort, the customs agent, and the customs supervisor, who had come over to see what the drama was about, vanished into a side room to discuss my fate. To this day, I have no idea what their conversation entailed. However, it must have been an epic saga. It took a full hour before they returned to me, and the escort handed me an official yellow sheet of paper with a good deal of Arabic writing and a stamp placed upon it. I would need this to leave Lebanon, he said. Lebanese customs had refused to put their stamp in any passport that had been desecrated by an Israeli visa, so they made up a whole separate sheet to stamp. The escort told me that, should anyone ask, I was a "merchant from America" who had "come to do business" in Lebanon. I can only assume that was the lie he had to make up to get me cleared to travel into the country. If just getting a foot in the door was this much drama, I dreaded what it would be like to actually work this case.

By the time we hit the car waiting to take us from the airport, it was 2:00 PM local time. The bomb had gone off over twenty-four hours earlier, and the scene was being held. There was no time for food or rest. The wicked could sleep when they were dead.

The FBI Legat whisked us away to the local police headquarters. Part of the investigation dance is to first interface with the officer in command. We were escorted into the large conference room, experienced the Offering of the Mini-Cokes, and got down to business. Briefings typically cover the event, and then the political sensitivities around the event. But this one had an odd vibe to it. By the time we got through with the formalities and briefings it was about three in the afternoon. The commander wanted to know how long we would be at the crime scene. Having no idea what shape the scene was in, I really had no solid answer. He seemed to be nervous about us being there much past five o'clock, but no reason was given. I should have picked up on that peculiarity.

We made plans to meet the commander the next day, after we had some time to review the scene, and took off to process the bombing. I had a very small team. In the past, I had traveled with more seasoned examiners; this time I was truly the team leader and on my own. The team consisted of two explosive specialists—one technician and one forensic chemist (Dave McCollam)—and one seasoned supervisor of the Evidence Response Team (ERT), which is the special FBI team dedicated to crime scene processing.

The FBI explosives chemists determine what explosive was utilized in the bomb. They sample for explosive residue, collect bulk explosive samples when they exist, and conduct the analysis of both. On a scene like this, I could have collected samples if necessary, but our chemists also serve another critical role. They interface with the local lab to ensure that the chemical analysis the lab is doing is being conducted in a proper fashion. If needed, they train local lab chemists in the specialized art of explosive-chemical analysis.

FBI explosive technicians are jacks-of-all-trades. Our technicians are not trained in forensic analysis, but they work enough bombings to know what needs to be done in such a scene. If things get complicated, examiners like myself make the hard calls. Our technicians support the gory, exhausting process of dissecting a bombing scene in too many ways to describe. Like nurses in a hospital, they are the unsung heroes of my world.

Our ERT personnel are trained to be experts in evidence recovery. Every field office in the FBI has at least one team of folks who volunteer to be trained in crime scene processing and evidence recovery. These are the people depicted in every crime TV show, those in the background with cameras and folks in raid jackets dusting for fingerprints, sketching rooms, and searching through desk drawers. They exist in real life, just without the mood lighting.

The goal at the crime scene is to collect what can be moved, and exploit on location what cannot. Full photographic documentation is a must. Bombs do damage to everything in close proximity. Documenting that damage helps to analyze the size of the bomb and lends clues to how it may have been assembled and functioned. Craters have to be measured and photographed. Damage to vehicles is likewise photographed and analyzed. The distance of broken windows from the bomb needs to be measured. Buildings nearby with broken windows need to have rooms searched to see if bomb components were thrown into them. Rooftops nearby need to be scoured for pieces of the bomb that might have launched that far. Photography would traditionally be handled by one person, measurements of blast damage by a team of two, and searches of rooms and rooftops by as many teams as we could muster. I had one guy to do all of these jobs. It was business as usual.

The bomb scene was a street. Kassir's demolished car sat next to a curb. The only thing blocking off the vehicle from the city itself was a flimsy ribbon of plastic crime scene tape strung in a circle, blocking off the street for about thirty feet in either direction from the car. The crime scene tape only blocked off the street. Sidewalks on both sides of the street were open for pedestrians and any casual observer to wander past. Search online for Kassir Assassination and multiple images crop up of me and my team processing the scene, with obvious observers watching from very close by.

We offloaded our massive black cases of processing gear and developed a game plan. Every scene is different. In the US, we have a well-established protocol we follow when processing any scene. In Lebanon, where our team was a tenth of the size needed to do the job, improvisation slowly edged out protocol. For our sole ERT guy, that meant working with our technician to grab only the most important measurements. The chemist (Dave) and I looked at each other, knowing what we had to do above all else, and started to process the car.

The epicenter of the blast in any bombing often becomes the epicenter of the search for evidence. As a rule, the first thing we do is collect trace explosive residue. As part of my research, I had examined a picture of the scene right after the blast. The street resembled a Black Friday sale. People mobbed the vehicle. Amongst the crowd, a wide variety of police and military uniforms were clearly visible. Military uniforms mean one thing to me. Contamination.

Police presence can contaminate an explosive scene. The gun sitting in an officer's holster traps Nitroglycerin released from the smokeless powder when a bullet is shot. If the officer touches his holster and then puts his hand on an object in a bombing scene, Nitroglycerin can be introduced to the scene. I had no doubt that bomb techs were present in the military personnel captured on film after the blast. It only makes sense to call in explosive ordnance disposal (EOD) after a bomb goes off. Military EOD folks spend their days surrounded by a wide variety of explosives. They may, or may not, wash their hands after using the toilet, but they rarely wash them after playing with high explosives.

I had to assume every surface of the car had potential explosive contamination associated with it. In a case where a foreign government is potentially responsible for the covert assassination of the journalistic voice opposing it, test results that show residue associated with military explosives could lend credence to preconceived notions on who was to blame. If we were going to find residue, it had to be trustworthy.

One of the first things that we do in a post-blast scene is identify the seat of the blast. That is the richest potential source of residue. In the car, it was not hard to put together a quick picture. The car floor was pushed upwards into the vehicle, the seat was heavily damaged, and the roof had multiple fragmentation holes from where the car floor, seat, and maybe additional fragmentation had launched upwards and outwards. The bomb was obviously placed on the underside of the vehicle. Had the bomb been under the seat instead, the car floorboards would have been blown downwards towards the road. I only had one problem. There seemed to be no crater.

A bomb that had done the kind of damage I was seeing to the car should have at least dented the street below it. The road under the vehicle was untouched. Old assumptions are hard to shake. In the US, the car would have remained untouched until after the whole scene was processed. Yet, here I was standing at what should have been the epicenter of the blast and the pristine condition of the road was impossible. My preconceptions acted like blinders. I called the Legat over and explained my confusion. My initial search for the seat of the blast to collect residue had been confounded by a lack of crater. Luckily the Legat had had some time to walk the broader scene and took me to a car parked behind the victim's car and, pointing to a hole under it, asked me, "A hole like this one?" So, either someone moved the crater, or the vehicles had been shifted. Applied physics ruled out the first scenario.

Who moves cars around on a crime scene? Was someone worried the time on the meter had expired? I still don't know why the vehicle was moved, but it was. We had our crater. We also had more evidence that the crime scene was potentially contaminated, perhaps worse than feared.

The crater did not look disturbed. I took some measurements of its size and then Dave and I removed the top layer of debris to get to the more sheltered soil beneath the road. When the bomb blast tore downwards and cut a divot in the concrete, it pushed residue along for the ride. The top layers might have been touched by those initially on the scene, but the layer accessed by a small shovel was still virgin territory. We got our first residue sample.

Collecting residue is something of a game of chance. Since residue splatters where physics decides to take it, collecting residue from only one area is like pulling a random card from a deck and hoping for a face card. The more draws you pull, the greater the odds of scoring your goal. Under normal circumstances in an untouched scene, we would have sampled the underside of the car immediately, as that was the area closest to the bomb. But in this case, as the car had obviously been moved, we weren't sure if it would be uncontaminated. We decided to look for a spot on the undercarriage that looked untouched and swab, despite the chances of contamination. It was too rich an area to totally ignore.

Next, we looked at the seat cushioning. The seat was ripped and mangled. Blood had soaked into the floor and seat where a portion of Kassir's legs were torn apart. All that cushioning was an ideal trap for particles of explosives. I dug into the cushion near an area of obvious scorching, while trying to avoid those that had acted as a blood sponge, to extract multiple sections of cushion for analysis.

As hard as we had tried, I was still not 100 percent convinced any of my samples were devoid of contamination. Then, as I looked over the scene, it hit me. The car roof. Multiple pieces of bomb and car had blown up from the vehicle through the roof. There was no point swabbing the exterior of the car roof as thousands of hands might have touched that. However, the inside was another matter. The car's headliner protected it. Dave and I cut out the headliner, which exposed the entire metal frame of the car roof from the inside. The only hands that could have touched this surface would have done so long ago in the factory that assembled it. We swabbed around all the holes where explosives traveled through, certain contamination did not lurk there.

With chemical forensics out of the way, it was time to move on to figuring out how the bomb was assembled and how it functioned. Measuring the damage to the surroundings helps determine the potential size of the main charge. I had already measured the crater, so I had some early indicators of size. Next, I took measurements of the hole in the floor under the front seat of the car. Knowing that makes it possible to judge dimensions of the device.

Explosives in contact with a surface create holes bigger than the charge, but never smaller. If a charge was in a Tupperware bowl four inches in diameter, I would not get a hole two inches in diameter. However, I might get one six inches in diameter as the metal stretched. Knowing the size of the hole and the distance from the bottom of the car to the street, I could get an approximate volume for the charge. A four-inch-diameter round hole in the floor of a car with a six-inch ground clearance gives a max size cylinder of a known size. Once volume is known, estimates of the type of density I would expect from the explosives found in the trace analysis give me maximum charge size. You can't cram five pounds of shit in a three-pound sack, but three pounds will fit rather nicely in the bag.

About the time we finished taking physical measurements and a full suite of pictures to document the damage to the car, the local police decided we looked like we could use some refreshment. As we walked over to our crime scene cases to work on a strategy for the next stage, a nice young officer walked up with two cans of Pepsi for Dave and myself. He was even nice enough to open them for us as he handed them.

There are a couple of very basic rules when working a crime scene. About at the top of that list is "no food or drinks in the crime scene." There is also a list of rules when working with a foreign country. On the top of *that* list is "don't insult the host by refusing gestures of kindness." Dave and I took the offerings but kept them next to our crime scene kit, which was about ten yards from the scene. Putting the Pepsis down next to the kit, we grabbed some more tools and headed back to the car.

Unbeknownst to either of us, a photograph of us pulling out our gear from the kit—complete with a prominent foreground display of the Pepsi cans, almost like an ad—ran in a news story picked up by the Associated Press. Later that night, as I called back to the lab to update my supervisor, my colleagues on the call asked whether we ordered in a pizza to go with our Pepsi. We were informed the picture had been posted on the web within hours of being taken, and that the entire FBI—even the director—had seen our unashamed shilling for the soda company. To this day, I am occasionally reminded of this forensic "gaff" by some of the older examiners with long memories.

But back to the forensics. Now we had to try to find pieces of the bomb. As I described earlier, every bomb has an explosive and a way of setting off that explosive. I was looking for the pieces of the mechanism that set the bomb off, and anything that might have been used to disguise or hold it. In my crime scene investigations, I often set up some possible scenarios to help guide such a search. In this case I came up with two.

In the first scenario, the bomb sat on the street under the car. Obviously, no one would place a bomb out in the open where anyone could see it. However, it was possible the bomber put the bomb in an old fast-food bag and placed it under the car. Someone casually driving by, or the car's owner coming from a vantage point where the underside of the car was visible, would easily discount the item as trash. No one wants to crawl under a car and recover some other person's refuse. There was a small crater on the ground, which meant a small charge could have been sitting on the ground. It could also mean a slightly larger charge was attached to the underside of the car, elevated a couple of inches off the ground.

In the second scenario, as in many assassination attacks, IEDs are secreted to the underside of a vehicle in a manner that hides them from open view, secured by magnet, glue, or tape. The decent-size hole in the car led me towards this possibility. With the bomb attached to the car, some portion will have to be directed upwards, whereas a bomb lying on the ground sends some pieces upwards, but the trajectory would also drive some portion outwards, like water pushed out sideways when a car hits a large puddle. I was not holding out much hope of picking up any pieces just lying on the ground of this particular bomb scene. I figured the local police and crowd of onlookers had taken care of them. My focus was on the car's interior.

While trying to figure out the best strategy to search the car, I turned back to my crime scene case to collect some more gear and stopped dead in my tracks. There, inside the wall of armed military officers and crime scene tape strung three feet off the ground that was intended to keep onlookers out, was a young boy about five years old, standing right next to my crime scene case. Perhaps he was a Pepsi fan, or perhaps he was just curious about the mysterious case and American men dressed in fancy protective clothing. Regardless, he should not have been in the middle of a guarded crime scene. I quickly pulled out my camera to capture the moment for posterity; it has become a favorite picture to use when talking about perimeter control. The boy was promptly removed from the scene at my request.

Finally, Dave and I began processing the car for pieces of the bomb. We swept out the floor and sifted through all the debris. Nothing obvious came from this process. Then we pulled out the seat cushion from the mangled seat frame and slowly cut into any pieces that looked like they had frag holes punched in them. We had just gotten into a good workflow when things went off the rails in a way I could never have seen coming.

When I first saw the orange shirts in the distance, I thought it odd but paid it no mind. Their matching orange attire made me think of some sports team, and I dismissed them entirely. The chemist and I were busy tearing apart the car. I had accessed the seat frame that would have been inaccessible before. I swabbed the inside of the car while the chemist crawled under the car like a mechanic to swab some of the harder-to-reach explosive holes.

The vehicle sat at the base of a quarter-mile-long hill. Traffic had been diverted from the road as the street was shut down to cars, but the occasional foot traffic passed by on the sidewalk.

Dave repositioned himself under the car, causing me to back away and glance up. Now, there were dozens of people in orange slowly making their way down the hill. More people merged into the orange mob from side streets, swelling the crowd to over a hundred. By now, they were a couple hundred yards away and moving with purpose towards us, descending the hill in total silence.

The mob continued to grow. Every inch of road was covered in a wave of slowly moving orange. When the human lava flow reached the crime scene tape, it parted like water around a rock and flowed around us on all sides. Within minutes, the entire crime scene was surrounded by the mob. Some climbed traffic lights to gain a better view of our isolated, fully engulfed island.

My partner came out from under the car. In the few minutes he took to focus on collecting residue, the world had changed.

"Hey, Dave," I mustered, staring at his shocked face, "we seem to have some company." He stood up beside me and, with no one there to explain what was going on, we simply stood stock-still and waited. The crowd stared back at us in silence.

Someone had to be the first to start it, but it seemed almost like a spontaneous upsurge, because the crowd suddenly erupted into song. Flags waved as hundreds of voices joined. Watching the security guards' relaxed body posture, I surmised it was the Lebanese national anthem. I took my hat off and stood at attention. My partner followed suit. I have no idea if it was the correct thing to do, but I needed to make some sort of gesture. As long as I did not have rocks thrown at me, I figured I was doing all right.

The Syrians had withdrawn from Lebanon approximately one month before Kassir's assassination after an upsurge of protests. This sea-of-orange protest had been organized to decry his assassination. The crowd had gathered to show Lebanese unity behind a silenced voice of resistance. Looking back, it became clear that the commander at the police station who pressed us on how much time we needed knew that a protest was to happen and failed to share that information with us.

After ten minutes of songs and chants, the crowd died down. Dave and I started to head back to the car and resumed our work. However, after about a minute the crowd became dead silent and parted, leaving an aisle. As the hush hit us, we turned back to the crowd to see coming down this aisle four policemen slowly escorting a woman across the crime tape and towards the car. I pulled Dave away from the car and removed my hat again. This had to be Kassir's widow. It was. When Kassir's wife got to the open car door we had just been working in, she slowly reached out a hand and, with just her fingertips, almost tenderly reached out and touched the perforated roof.

I had no words. And I still have no words. She was sorrow incarnate. A gentleness touching atrocity as if hoping contact would somehow bring clarity to such a brutal act. I am lucky; as a scientist, I don't have much contact with victims. However, this moment remains galvanized in my memory as emblematic of all those who have been hurt by the acts I help investigate, of those struggling to comprehend such tragedy. I cannot bring spiritual relief, but, hopefully, I can play a role in bringing some justice.

The widow was slowly escorted away, and the crowd dissipated back into the cityscape from which it had materialized. I was left drained as the adrenaline faded like the press of orange that had so quickly dissipated. Back to business, plans were made to haul the car away to an impound lot so we could work in a more secure environment. Daylight was passing quickly, and a new time pressure was upon us. We had to get to the morgue that evening. There was a dead body awaiting my attention.

In theory, a deceased Muslim should be buried as soon as possible. Many believers try to have the burial take place within twenty-four hours of the person's passing. Kassir's body had been lying in police custody longer than that. The funeral was to take place the next day, and if I wanted to examine the body, I had to do it that evening or not at all.

By this point, I was running on thirty hours with little sleep and the better part of a day with no food. The caffeine from the crime scene Pepsi was losing its edge. But the day pressed on.

Morgues, as you now know, have a special place in my deployment experience. Mercifully, the bomb in this case was a surgical strike. I would not be looking at miscellaneous body parts. But that didn't make the experience any less unsettling.

When we got to the morgue, Kassir's body, wrapped in multiple layers of plastic sheeting, was laid on a metal table about four feet off the ground. We debated whether it was best to slowly pull the sheeting off or rip it off as quickly as possible, like a Band-Aid, but neither option would be pleasant. I knew the visceral reaction of immediate confrontation with maimed flesh was a bit of a mind smack. The body wrapped in plastic on the table offered another type of disconcerting experience.

As each layer of plastic sheeting was removed from the body, physical features slowly revealed themselves. Other features, like the vague red mass, remained puzzling. When the last plastic sheeting layer was removed, it dawned on us: the vague red mass was a hunk of flesh taken out of the hip.

Strange thoughts go through your head in times like this. It occurred to me that this was the first dead body I had examined that still had a head attached. I could not parse whether that made the experience better or worse. There is a disconnect when looking at body parts, headless corpses, and severed heads. In some way, they seem less real, albeit fairly grisly. Kassir was still ostensibly intact, and that imbued the body with more realism. As the bomb had been placed under his car, the majority of the damage was to his lower limbs, hips, buttocks, and thighs. With major arteries severed, he bled out fairly quickly.

It is possible to get explosive trace residue off a body, but I did not feel it was necessary with the processing we had done of the car earlier. The body also looked like it had already been partially washed. We turned the body to the side so I could get a full assessment of his injuries, while my team took pictures of the physical damage. I used this as part of my estimate for how big the bomb was. I also looked for fragment strikes while moving the body. Bombers often pack their devices with a variety of added items, like nuts, bolts, and screws. When they penetrate into the body, they leave distinct markings. Nails and screws can leave long lines, gashes, and holes. Ball bearings leave only holes. Pieces of a car floorboard come apart in irregular patterns and, as such, leave perforations of a variety of sizes and shapes.

I saw no obvious pattern of frag wounds on the body. I started to wonder if any fragments from the bomb had been imbedded in the body. In my earlier processing of the car, I had found nothing that appeared like part of a bomb fuzing system. This struck me as odd, but since I had not had a chance to finish my work before the protest, I was not sure what later analysis would recover.

I asked if the body had been x-rayed. It had not. It is not routine procedure to x-ray a body in a morgue; however, X-rays can show metal fragments trapped in a body, and we have seen this in numerous scenes over the years. When I explained about the bomb fragments, the coroner nodded in understanding, walked to a side table, and picked up something. He then handed me a scalpel and took a step back, motioning "have at it" towards the body. So there I found myself standing dead still glancing alternately between the surgical instrument in my hand and the smiling medical examiner beaming with pride at his willingness to help our investigation. I turned back and saw the look of bemusement on the faces of my team members. To a man I know they were all thinking, "Thank God I'm not the one who asked that dumb-ass X-ray question."

I have done many distasteful things as a forensic examiner. I was not about to add desecration of a corpse to that list. Slowly walking up to the body, I decided to pull out my magnifying glass and take some time to closely examine a few wounds. Nodding a couple times, and shaking my head at others, I put the scalpel down. We thanked the local police and medical team for their time and headed out. On the car ride back to our accommodations, my colleagues offered up a few snide comments about my acumen as a surgeon.

After the long flight, the customs conflict, the meetings with local police, the attempts to process the crime scene, the mob sing along, and an unaccepted offer to attempt human dissection, I was tired. It would have been nice to just head to bed, but that wasn't going to happen. During this time of political turmoil, there was no safe hotel for our team of Americans in the city, so we were housed in rooms around the US Embassy. Before I got any rest, I had to finish up my crime scene notes.

Memory is fragile. Reliance on it should be kept to a bare minimum even when fully rested and faced with minimal distractions. That is why the first thing I did when I got into my new temporary abode was break out a notebook and make a sketch of the general crime scene. While surveying the car, I had taken numerous pictures. I had also taken pictures of damage to the surrounding buildings from the vantage point of the vehicle. With so many pictures, it is sometimes hard to remember what each depicts. A true crime scene photographer sketches the scene being photographed and notes on the sketch where each photo is taken and what each photo depicts. Luckily, digital cameras allow for quick review of all photos.

I sat in the small apartment's living room table and spent an hour reviewing each photo and noting the details of each on a sketch I made.

This done, I decided to jump in the shower to wash the dust, residue, blood, and grime off of me before finally falling into sleep's abyss. Showering after working a crime scene becomes almost as much a mental cleansing ritual as a physical scouring off of the myriad unpleasant things that follow you from a scene. I was just starting to feel the stress drain away when I looked over to the shower curtain. There at eye level, about one foot away, was the biggest goddamn spider I had ever seen in my life. In general, I don't suffer from arachnophobia. But having a tarantula shoved in your face while showering transcends normal conditions.

Fear is a product of millions of years of evolution distilled into actions and reactions designed to ensure survival of a biological entity. The neural circuitry of fear starts in the amygdala within the brain. A brain continually processes stimuli from the optic nerve to allow us to get about our daily business. Light captured by my eyes reflected by the spider triggers the optic nerve to pass along a series of signals that the brain interprets as an arachnid bigger than a dinner plate. At that point, the amygdala hijacks the brain and circumvents the "thinking" part housed in the neocortex. The amygdala triggers the hypothalamus, which produces the hormones dopamine and corticotropin, followed by an activation of the sympathetic nervous system. Corticotropin further activates the pituitary gland to release the adrenocorticotropic hormone (ACTH), which then stimulates the adrenal glands to produce cortisol and adrenaline.

A body deprived of sleep, and already adrenaline-fatigued from the earlier mob scene, reacts poorly to more stress hormones and adrenaline pumped into the system. In short, the rational part of the brain checks out and wishes the best to the other regions at the "fight or flight" helm.

I don't recall how I got there, but in the next instant I was out of the shower. When my heart slowed down enough for me to hear the falling water in the shower again and I was able to regain control from the amygdala, I plotted the best way to dispose of my shower mate. Thank God the FBI has never seen fit to arm scientists because I would have given serious consideration to emptying a clip into the eight-legged intruder. Absent projectile weaponry, I slammed two boots through the shower curtain with the force more applicable to taking down a wild boar. Foe vanquished, I decided I had had more than enough for the day and it was time for bed.

The new day brought new challenges. There was some speculation that the bomb was set off by a mysterious motorcyclist who was parked at the top of the street where the mob emerged the day before. We were asked if we could review security tapes from a nearby mall to see if we could spot the rider. On the face of it, this was not my job. Investigative agents on the ground run leads like this. However, we had a small team, and the Legat wanted me along to see if I saw anything that I could identify as a bomb trigger—if we were able to locate the tape. With the morning free, I saw no reason not to head off on an adventure.

As we parked at the mall's underground parking garage and headed to the security control center, I noticed a line of cars entering the mall being stopped and examined by mall security. When I saw how the cars were being searched, I froze in disbelief. This was the first time I had come face-to-face with one of the biggest explosives scams in my field's history. I was seeing the mythical "explosives divining rod" for the first time in person.

The story of the divining rod could take up a whole chapter on its own. It seems I have been fighting this scourge for almost as long as I have been with the FBI. Although you might not think it, as important as my function is to bring science to bear in the pursuit of justice, an equal part of my job is to shine the light on scams and junk science. There has been no greater scam in my career than the explosives divining rod, which we called EDRs. Incredibly, it all started off with a golf ball detector. The root of EDRs dates back to 1993. A former used car salesman invented a device to find lost golf balls, called the Gopher.[16] Later, the used car salesman remodeled the Gopher, removed the label, and applied an explosives detector label. He named this advanced explosive detector the Quadro Tracker Positive Molecular Locator ("Quadro Tracker" for short).

The used car salesman did not skimp on the claims he made surrounding the Quadro Tracker. Amongst the items the device was touted as being able to find were explosives, drugs, weapons, and currency. Although I couldn't track down original brochures for the Tracker, I located newspaper articles from the time period that all note that the device was purported to function by oscillating "static electricity produced by the body inhaling and exhaling gases into and out of the lung cavity" to "charge the free-floating neutral electrons of the signature card with the major strength of the signal." These statements utilize a broad range of scientific terms with as much validity as a mad-lib, but they fooled the non-technical audience. The device was constructed out of a plastic pistol grip with a retractable car radio antenna attached to its front on a swivel. The idea was to walk around the vicinity of a suspected bomb and, when the antenna pointed at something, it contained explosives. Approximately one thousand of these devices were sold to numerous school districts and public safety departments across the United States. The FBI conducted testing of the devices with Sandia Labs and concluded "the advertised mechanisms of the Quadro Tracker device appeared to not be based on any known laws of physics." In other words, the Quadro Tracker claims were full of shit.

Eventually legal action would halt sale of the Quadro Tracker, but the genie had left the bottle and other scam artists would continue to market a new generation of miracle detectors based

[16] A couple years ago, I purchased one of these magic items for twenty dollars. It still sits proudly on a bookshelf in my office.

on the same principle. Such phony detectors would spur a series of new scams, further knockoffs, and continual criminal investigations that would go on to the present day. The Quadro Tracker would go on to morph into the MOLE, ADE651, GT200, and Alpha 6. Some of these scam detectors would sell for as much as $40,000 each. Following the Gulf War, Iraq would purchase upwards of six thousand detectors for $60 million. Apparently, they got a bulk discount. Bomb squads not just in Iraq, but also in a wide array of foreign countries with little to no understanding of the product's limitations purchased an estimated several thousand of these units.

And here in this parking garage was one of these POS detectors out in the wild, a waste of thousands of dollars doing nothing.

Unfortunately, the video review turned out to be fruitless. This was getting frustrating. Not only was I finding no clues, I was stumbling across stuff that was inconceivably wrong that I needed to correct totally unrelated to this case.

Meanwhile, the local police were working on getting me access to a car dealership. While the prices of motor vehicles were cheaper in Beirut than my home state of Virginia, my interest was not consumer. When examining damage from a vehicle bomb, it is necessary to know what the vehicle—in this case, an Alfa Romeo—looked like before a huge hole was torn into it and blast forces rearranged its interior.

I spent about an hour at the dealership examining the exact make and model of Kassir's car and taking pictures of the wiring underneath the dashboard and the construction and mounts of the seats. Bombs are often comprised of wires. After the bomb went off under Kassir's car, the blast wave ripped a hole in the floor and tore up all the wiring nearby. There is no way to tell which wire goes with a bomb and which with the car unless you have an exemplar to look at. The same goes for electronic components, hardware, and screws. Photographs of all the components in an intact car help me distinguish between IED and vehicle.

We still had to determine if the bomb was placed on the ground or attached to the undercarriage of the car. With the vehicle mounted on a mechanics lift and raised up, I could examine the entire undercarriage to determine where potential spots to attach a magnet might exist. The vehicle possessed what is known as a unibody frame and there were very few places anything magnetic would stick. This, coupled with the lack of tape found on the scene, led me to believe the device was disguised in some sort of paper bag placed under the car.

With a full catalog of the vehicle made, I headed off to the impound lot where the car had been taken so I could finish the exams I started—and had interrupted—the day before. By this time, a French team had arrived on scene, and we had made arrangements to meet up together at the impound lot.

It is common to have multiple teams on the ground at these sites. My goal was to finish tearing down the victim's car. Luckily, I had worked with the French team in the past, and we knew each other. Our teams systematically pulled every seat out of the car. The great thing about being in the impound lot was that I had access to the mechanics and all their tools. We unbolted and removed every single seat from the car. Trying to work within the confines of the ruined husk of a bombed-out car requires flexibility more akin to a contortionist than someone whose limberness had long since passed the point of being able to touch toes. Having seats pulled out and sitting on the pavement made for fewer necessary yoga postures. In addition, my team's expanded numbers made processing much faster.

Typically, in such a scene the floorboards of the car yield some evidence right away. That had not happened on my first day. With nothing on the floorboards, we checked the seat cushion under the victim. That also yielded zilch. A feeling of unease came over me. I had never worked a scene and not found evidence. Our teams were finding nothing. The more we found nothing, the stronger my determination became to tear the car apart bolt by bolt. We ripped through every bit of cushioning in the car. We ran magnets over every piece of debris. We tore out carpeting, ripped apart the dashboard, and flipped the car on its side for full access to the undercarriage.

I went over every inch of the exterior and interior with a magnifying glass to find remnants of wire, tape, magnets, batteries, electronic components, or any of the myriad items typically utilized to make a bomb. Either the bomb was set off by an action of the victim, or a command from the bomber. Both of these scenarios required a switch of some kind. If victim-activated, the switch could be as simple as a clothespin. If command-detonated, electronics would be involved. Both our teams worked hours and found nothing.

It is difficult to explain the frustration this created. Post-blast investigation is predicated on some very simple tenets. Number one: bombs are made of stuff. Number two: when bombs go off, stuff gets thrown around. Number three: investigators need to look for stuff that looks like it has been through a shredder while set on fire and collect that stuff. Having nothing to collect was challenging tenets one and two. So, either the bomb was a bare explosive that magically went off, the materials utilized to produce the bomb self-destructed like the *Mission: Impossible* instruction tapes, or someone else collected everything before we got to the scene.

The most likely scenario was that someone had collected pieces of evidence before we got to scene. But I had seen the state of the car that first day. No one had thoroughly processed the car. The seat cushions were intact and the dashboard was still in place. It is possible someone picked up obvious fragments just sitting on the floor of the car, but things would have been trapped in many other places. I should have found something. As part of my job duties, I had spent years building small bombs and putting them in all sorts of locations in vehicles. After blowing these bombs up, I led teams of bomb techs being trained in post-blast to process the cars. I knew where things got lodged. In Kassir's car, there was nothing there. It was really pissing me off.

After a couple of hours, we had to give up. I had never worked a scene like this. The French team was baffled as well. The French wanted to collect some of their own swabs before we departed. Most likely, they decided to do this so they would not leave totally empty-handed. We agreed to share chemistry results with them after our analysis.

After leaving the impound lot, we made a quick visit to the local crime lab. In some countries, this is a two-part visit. In the first part, I talk to the chemists and visit their facility to see how they analyze. Then, I meet with the local bomb techs to get a better understanding of their capabilities. As we headed to the lab, the embassy staff informed me that politics were pretty tense in Beirut. In fact, the police and the military were in direct conflict with each other. I would not be able to meet with the military, as I was helping the police investigation. That may sound insane because it is insane. But rivalries between civilian law enforcement and military units are commonplace worldwide. Many countries focus all resources for explosives and bombs within their military EOD units; these countries can't afford to train the civilian police force to deal with this threat. Unfortunately, the military units are also not provided training on how to process a crime scene. This leads to some serious disconnects.

There was nothing to really see at the local crime lab, which was far less than state of the art. The lab was not equipped or trained to deal with specialty cases such as bombings. Lebanon had recently been through a vicious civil war, followed by internal struggles with Syria, and therefore had limited resources. Beirut's war-torn history was vividly displayed throughout the city, with multistory concrete buildings scarred by shell impacts and bomb blasts, and its interior infrastructure suffering. Carried forward on a tide of tragedy, building a forensic crime lab could not be expected to rise to the top of priorities for this harried nation.

After the crime lab visit, we headed back to the embassy. Being part of an FBI team means no end to the deployment meetings. While my days were filled with processing crime scenes and developing some semblance of a forensic picture, evenings were occupied with preparing update reports for the FBI Mothership back in DC. The US ambassador wanted a briefing from me that evening. Also, the Lebanese general in charge of the investigation wanted to meet with our team the next morning before we headed out of the country. Both of these were no surprise, but they added up to create a sixteen-hour workday.

Our team went out for supper down in the oceanside part of town. Our embassy staff knew of a nice fish restaurant that overlooked the sea. The weather was warm, and a breeze blew in from the water. We sat on an outside patio and enjoyed the scenery for the first time in our trip. The restaurant had fish tanks distributed around where you could pick out the type of fish you wanted to have prepared for the evening meal. If one particular fish was giving you the stink eye you could pick him out for retribution. None of the marine creatures on display seemed to hold any acrimonious attitude towards me, so I let the embassy staff choose our entries. We got to enjoy a brief respite of a couple hours to unwind before I had to brief the ambassador. Along with the fish we engaged in the local tradition of an after-dinner smoke with apple wood shisha charcoal. Most people are familiar with this due to the evolution of trendy hookah bars, but the tobacco scented with the sweet and fruity apple wood drawn through a water pipe does make for a soothing repast. This coupled with a small glass of arak, an anise-flavored liquor mixed with a small bit of water to induce a slight haze, makes a fine finish to a day. Had I not had a scheduled meeting with the US ambassador I would have sunk a little deeper into the arak, but I had enough to dispel some of the tension from the last couple of days.

When I got back to the embassy the rest of the team headed back to their rooms to get some rest, and I headed to the ambassador's residence with the Legat. Ambassadors run a wide gamut of personality types. As political appointees, they may or may not have any real insight into the culture of that particular country.

Most are very social; few understand anything scientific. This means part of my job is to educate them on bombings and how the FBI investigates a bombing scene. I spend a great deal of time explaining the evidence I found and how it fits together to make a bomb. In this case I found nothing, which was baffling me, so explaining this to an ambassador would be doubly challenging.

I brought with me some PowerPoint presentations I always travel with for teaching purposes and met the ambassador at his house. It was about nine at night, so he was at home with his wife enjoying an after-dinner drink. We retreated to his study so I could show him some computer slides about explosives and bombs, as well as show him some of the pictures I had taken of the scene. I had done this type of briefing numerous times, but never with the ambassador's spouse present. I decided not to show any pictures from the morgue, as once you get those in your subconscious, they set up tent stakes and don't go home. It seems that my choice to shield the wife from unnecessary gore was prudent. At one point during my talk, I showed a video of a bomb about the size of the one used against Kassir's car to give the ambassador an appreciation for the power of such a device. When the bomb in the video went off, the ambassador's wife was so surprised that she screamed and rolled back off the beanbag chair she was perched upon in front of the computer monitor.

I apologetically helped her back to her feet while my brain started to compose the email I'd have to send to the FBI director about how I scared the shit out of the wife of a lead diplomat. Thankfully, the ambassador was slightly embarrassed by the almost-slapstick theatrical display and brushed off my concerns. It was the perfect end to my carnival ride of a day.

The next morning, I prepared myself for our team meeting with the Lebanese officials. Interfacing with high-ranking government officials and political intrigue comes without a training manual. They don't provide a block of instruction on that when you are taught how to be a forensic examiner. To be honest, scientists are drawn to the lure of isolated labs and long hours of bench-top focus for a reason. We are not built for cocktail parties, small talk, and parsing of every word in an attempt to avoid an international incident. But I was on deck, and all I could do was get ready to swing the bat.

I was ready for a technical meeting with the head of the Beirut police. In addition, I was prepared for the inclusion of the general himself, who oversaw the entire police division. I did not, however, expect the paparazzi. When we walked into the large conference room for the meeting, we were greeted by half a dozen flashing cameras and a massive shoulder-mounted news camera, like you see on TV. I was introduced to the chief of police. Cameras flashed; photographers ducked low for the dramatic shot. The general shook my hand, the video camera spun to the side while more flashes went off. By this point in my career, I had gotten used to readjusting expectations and rolling with the moment. OK, the meeting had turned into the Oscars' red carpet. Got it. Who was I wearing? Why, this is a little number put together by Royal Robbins.

So why the cameras? It took me a few minutes to parse what was going on. I realized that the case was huge news, but I did not realize the political implications it had. When the FBI arrives at a crime scene, there is a mystique we bring with us. I have seen the wizard behind the curtain, but all the Lebanese saw was the great and powerful Oz. Beirut authorities wanted to ensure the public—who had gone through months of riots and protests—that the government in charge was serious about solving this case. What would show that better than dozens of pictures of FBI bomb investigators meeting with the Lebanese police? I am sure my picture was all over the local papers that night, and video splashed across all the news stations. I hope they caught my good side.

After the photographers had fully documented the FBI team and our sincere handshakes with the local police, all photo-documentation teams departed the meeting and real work ensued. I gave a broad assessment of what happened. Enough physical evidence of damage was present for me to hypothesize a couple of scenarios. I promised a complete official FBI report would be submitted based on my analyses very shortly after I returned home. Finally, I addressed the lack of physical evidence related to the bomb. I believed it was unusual and noted that we remained stumped by the situation.

The general asked for an assessment of his lab and capabilities. This is where things always get tricky. Under no circumstance will I ever speak in derogatory terms about local teams. Nothing good can come from that. Instead, I explained how bombings were very difficult to investigate, and how even the FBI is challenged by them. Explaining how the FBI had investigated bombings since the 1920s allowed me to explain how we did not start off as the best. We learned from mistakes and developed the procedures we use today. I didn't want bruised egos or developed defensiveness. I offered to develop a training plan for bomb techs and investigators in Beirut. The FBI would train anyone the general wanted on the way we have learned to process a scene. We would partner with them to help develop lab capabilities. We would assist in future cases if they so desired.

This meeting turned out to be the most important part of the trip. Out of this offer grew a partnership and the building of very valuable liaisons. I and my colleagues in the Explosives Unit taught the first team of Lebanese bomb techs in an FBI "Post-Blast" class. This built strong, lasting bonds between our teams. Years later they would ask for our assistance after 2,700 tons of Ammonium Nitrate stored in a warehouse detonated after catching fire and devastated a huge swath of downtown Beirut.

Kassir's assassination was not the end to the saga of Lebanese unrest.

Following his murder, multiple bombings occurred in Lebanon, including the George Hawi assassination and May Chidiac attack via IEDs targeting their vehicles. The FBI sent teams to both of these events, as well, and also found little evidence, as I did.

From a forensic perspective, the two deployments that followed mine offered some vindication of my team's results. When I debriefed my fellow examiners about the case, I caught some "professional" critiques about not finding any evidence. No one in the unit had ever been to a scene and not found anything. But when the next two teams returned finding exactly as little evidence as I found, some serious rethinking occurred.

Whereas in my situation, we felt that some of the evidence had been compromised by the lack of scene control, I was still unable to explain how my tearing apart the car to its component parts yielded nothing. In the subsequent deployment, the agent had a military escort and the military had sealed up the entire scene, refusing to let anyone near it until the FBI team arrived. He still found nothing. Something bigger was going on.

If you are going to aggravate someone to the point where they start plotting your murder, it is best to ensure that the person who wants you dead isn't a medical examiner. Examiners probe into causes of death for a living and can most likely devise something that would be either impossible to detect or so obscure no one would think of looking for it. In the same way folks like me can design bombs that would be very, very hard to figure out. Whoever was assassinating highly visible people in Lebanon was a professional. That is about all I can speculate on regarding this issue. It was the only time in my career, and the history of the FBI Explosives Unit, that we had been stymied. Perhaps President Assad and his intelligence apparatus were not as far away from the Lebanese capital as we thought.

CHAPTER 8

Going to the Dogs

In March of 2018, Austin residents lived in abject fear for nineteen days as one bomb after another, each basic in design but lethal in effect, exploded, killing two residents and injuring five more around the central Texas town. Though the bomber, who shipped his lethal deliveries under the mocking alias "Kelly Killmore," lived and walked among the citizens he terrorized, tracking him—like tracking so many spree and serial bombers before him—was akin to connecting dots drawn in invisible ink.

As a forensic scientist, I've mastered the tools needed to determine what makes a bomb "tick." I know how chromatography is used to analyze the finest traces of post-blast residue to tell you precisely what materials a bomber used to build and detonate his destructive device. What I don't possess and have been relentlessly pursuing with my peers are tools that make it easier to pinpoint the actual bomber—especially when a city (or nation) is living in terror while random people in their community are being killed and maimed. Before he died in 2016, a colleague and friend almost convinced me that this tool existed right under our noses—or, more specifically, under our canine companions' noses.

Anyone who's ever been greeted with a wet nose to their packed bags in the airport security line knows that we already have bomb-sniffing dogs. They're very effective, because that job requires skill akin to that of a sommelier. Just as there are only so many kinds of grapes, there are only so many explosive compounds in the world. If I want a dog to find TNT, for example, I can have the animal trained with a real explosive. From that point on when the dog picks up the scent of TNT, it can let us know.

Finding a person who has built a bomb is more challenging, but it's also a job that certain dogs could be trained to do. Every living human emits their own unique odor, like a fingerprint for the olfactory senses. We can't sense it. But human scent dogs—bloodhounds—can.

When a team of scientists published a 2004 paper titled *Survivability of Human Scent,* showing that dogs could pull an individual human scent from a mass of putrid, acrid, explosive wreckage, some forensic experts believed we had uncovered a game changer—that we could literally unleash not only a bomb-sniffer but also a bomber-tracker into a crime scene to stop a spree before it starts.

My foray into human scent tracking was the only time in my long career with the FBI that I was nearly lured away from explosives. Little did I know where those dogs were about to lead me.

Like most people, my first exposure to bloodhounds was pretty much limited to TV and movies: Someone breaks from the chain gang, the Southern prison guard with his Ray-Bans and toothpick calls for the "hound," and the chase ensues. In the prison-break scenario, the task doesn't seem too hard. The dog chases someone's scent through woods, fields, and streams. Not much other human odor is around to confuse things. But the notion that these animals could rummage through the chaotic olfactory wreckage of an elusive bomber, combined with possibly multiple victims and myriad law enforcement officers, seemed well beyond the scope of those Hollywood portrayals.

Yet, the stories that emerged from forensic teams who used canines to track down bombers were practically mind-blowing (so to speak). In one case, a nail-and-gunpowder pipe bomb exploded inside a Chevy Blazer in Washington, DC. The explosion scorched away a large portion of the driver's skin and tore the buttock off one of his legs. Three days later, the driver's half brother, Prescott Sigmund, went MIA. Authorities found his car in a Vienna, Virginia, metro parking garage with what seemed to be a suicide note.

Seventeen days after the bombing, a dog used scent pulled from the pipe bomb fragments to take agents from the Bureau of Alcohol, Tobacco, Firearms and Explosives (ATF) and the FBI's Human Scent Evidence Team (HSET) through an unknown neighborhood directly to Sigmund's house and then basically plopped his furry butt down right on Prescott's front doorstep.

Following the alert at the house, the dog later led the authorities from where the car was abandoned at the parking garage, through a crowd of commuters (not to mention a sea of human scent left behind after two and a half weeks), to the elevator and, eventually, directly to the bus stop Prescott used to leave town. (Prescott ultimately turned himself in after being featured on *America's Most Wanted.*)

In my world, no tool even comes close to that. It could literally take years to do what that dog did in just the better part of a day. A canine's nose houses up to three hundred million olfactory receptors compared to our measly six million, and their olfactory processing center is, proportionately speaking, more than forty times larger than ours. Those animals see the world through scent. And nobody believed in their potential for forensic work more fervently than Rex, who, despite his somewhat ironic name, was not a dog.

Rex Stockham,[17] an FBI explosives and hazardous device examiner, was the main point of contact for all our bloodhound work. He became hooked on the hounds' potential at a conference presentation from a group of bloodhound handlers from California about some experiments they were doing with drive-by shootings.

Everyone has heard about drive-by shootings. Without witnesses, such crimes can be tremendously difficult to solve. Often, the only piece of evidence is a shell casing from a bullet that gets ejected from the gun as it is shot—that's not a lot to go on.

One day on the range, the California bloodhound handlers wondered if they could get enough scent off a single bullet for their dogs to track from it. They had someone shoot a bullet and then run off to see if the dogs could follow his trail. The teams presented odor from the bullet to their dogs, and, in return, were greeted by blank stares and a lack of enthusiasm. It seemed the answer was no.

Undeterred, they figured maybe one bullet wasn't sufficient, that maybe there was some minimum number of shell casings the dogs needed to track the odor. So, their pretend perpetrator loaded and shot a batch of fifty bullets, and then ran away. When the dogs were exposed to the scent from this larger batch of bullets, they were able to discern the person's odor and track

[17] After becoming close friends, Rex would tell me the story about how he attempted to keep me out of the FBI after reading my application package and concluding I was "nuts." Rex passed away from World Trade Center–related cancer in 2016. I have yet to meet anyone who exceeded Rex's passion for the FBI mission.

him. The handlers' hypothesis was correct; it was only a matter of quantity.

Over the next few months, the handlers trained their dogs with the odor from fewer and fewer bullets. Within a month, some dogs could track from the odor left on only twenty bullets. At that point, some dogs stalled out, but others kept getting better. When the entire process reached its end, the handlers had three dogs that were capable of tracking a person based on the odor pulled from a single bullet.

That solved one half of the problem. The other part of the dilemma was the "drive-by" part of the equation. Top trained bloodhounds might be able to get someone's scent from only one bullet, but the person shooting is in a moving vehicle. When you run down the street, you throw off odor. In the fugitive example, the escaped prisoner brushes past leaves and long grass, leaving behind skin cells and odor. When that same escaped prisoner runs on a street, gravity ensures that he makes pretty regular contact with the ground. His scent, which adheres to the soles of his shoes, is the olfactory equivalent of a trail of footprints in paint. But in a car?

I soon learned your car is basically a big bucket of your scent. As you drive, you grind your scent into the upholstery, seat cushions, and floorboards. Cars are not hermetically sealed, so, therefore, the occupant leaves a spray-painted trail as the car spews his scent down the street behind them. Amazingly, bloodhounds can discriminate between the scent of the perpetrator and all the other scents.

Upon hearing the story of the bloodhounds' success with drive-by shootings, Rex envisioned what they might be able to do to help bomb cases. A person putting a bullet in a gun needs to spend very little time in contact with the bullet. However, building a bomb is a time intensive action. There is a good deal of hands-on activity in the construction of IEDs. This construction typically takes place in a location familiar to the bomb builder, which is drenched in his odor. If dogs could take scent off a single cartridge casing, an IED should be much easier. The only problem is that bombs, on occasion, do blow up.

At the time, no one knew what effect an explosion would have on the scent left behind on the bomb fragments. Explosions can generate tremendous amounts of heat, which tends to drive off volatile chemicals. It would take some research to determine if hounds could still trail from the scent of charred debris. That's where I came in.

Rex teamed with the Southern California Bloodhound Handlers Coalition and built simple IEDs, like pipe bombs, then blew them up. He constructed arson devices that burned for two minutes before being extinguished. He pulled the scent from the recovered debris, such as a gas can, a piece of pipe, a section of a battery, or the fragments of a switch. The dogs were able to take scent off all of those and track the builder. The technique had promise. With a simple proof-of-concept behind him, Rex turned to me for a deeper dive.

Together we made it trickier. In one test series, I built an eleven-pound explosive charge using Hydrogen Peroxide, a very popular (and unstable) material with terrorists. The pretend bomber sat in a pickup truck and handled the steering wheel. Then, I placed the bomb in the passenger side floorboards and set it off.

It is hard to explain what eleven pounds of explosive can do to a vehicle. The photos on page 134 are two frames from a video we made of the explosion. In the top photo, the truck's rear bed is visible under the fireball. In the bottom photo, I am posing with my handiwork. You can see the remains of the steering wheel in this picture. We were able to pull the pretend bomber's scent from this wheel, which the bloodhounds successfully trailed. The peroxide explosion hadn't "sterilized" the scent, as I'd feared.

Emboldened by these research tests, our next step was to try to flex our muscles on some actual cases. And before long, the perfect case came our way.

It all began with mailboxes, among the most clichéd bombing targets, especially in the hands of simple-minded juveniles armed with pyrotechnic devices. I'd witnessed my share of these long before the notion of a career as a bomb detective had ever taken hold.

One night when I was just fifteen or sixteen, a good friend and I were sitting around playing on my Atari game system when our fight with aliens was interrupted by the most thunderous explosion I had ever heard. Preceding this blast was a loud screeching of tires. We both ran out the front door to see my family's mailbox blown across the yard and smell the acrid odor of burning flash powder.

My father was an English teacher at the local high school, located right across the street. He was also a hard-ass who stubbornly tried to reach those who had little interest in learning. Our mailbox suffered for both my dad's expectations of effort from students and our proximity to the scene of the crime. In those days, it was still possible to get the classic M-80 firework filled with a three-gram charge. Many a mailbox, including ours, fell prey to this device, as did numerous digits of kids who failed to release them in time. The latter factor eventually led to the fireworks being outlawed.

My father eventually built his own mailbox with a Plexiglas window on the back face so that he could see any suspicious device ahead of time. Had others had the same mailbox design as my dad created, they might have been saved from injury during the bombing spree of Lucas Helder.

Lucas's misadventures started on May 3, 2002, with five pipe bombs placed in residential mailboxes in eastern Iowa and three more devices rigged in mailboxes in an identical fashion in western Illinois. The explosive devices were soon scattered across the country like shrapnel, with another eight pipe bombs discovered in mailboxes in east central Nebraska, then two more recovered in mailboxes in Salida, Colorado, and Amarillo, Texas. Unlike the first round of devices, the second round did not have complete fuzing systems, so thankfully they never exploded.[18] By the time the bomb spree was over, the authorities documented a total of eighteen pipe bombs.

At the time of the attacks, I was finishing up my training as a forensic examiner in the Explosives Unit. While I was back in the lab examining the bombs being sent in, Rex rallied the bloodhound teams to see what insight they could provide.

This was not as simple as trailing a car in the wake of a drive-by shooting. Typically, such shootings occur within miles of the shooter's house. Lucas Helder's bomb placements, by contrast, spanned multiple states and his route covered approximately 3,200 miles. It was not practical to run bloodhounds from the scene of any of the bombings. So, instead, the teams took scent from the most recent vandalized mailboxes and let the hounds start tracking.

Western states have vast tracts of open land. A road can go on for miles before a turnoff or intersection. There were only two options for the bomber when placing his devices. He could drive up to the mailbox, place the bomb, and keep driving in the direction he was heading, or he could drive up, place the bomb, and turn back around and return from the direction he came.

Bloodhounds always track the freshest scent. The team drove to the most recent mailbox and let out the hounds, who started trailing in the direction the bomber last went. Then they'd pile the dogs back into the cars and drive to the next decision point in the road. If the road led to a four-way intersection, the teams got back out, took the scent, and set off the dogs, who trailed either

[18] The devices were relatively simple in design and consisted of one-inch-by-six-inch steel pipes sealed with end caps containing smokeless powder as the filler. Each pipe utilized a light bulb as an initiator, a nine-volt battery to light the bulb, BBs and/or nails for added fragmentation, and a switch consisting of a piece of string tied to a spring designed to make contact with a paper clip. They were rigged in such a manner that opening the mailbox door would close a switch and set them off.

left, right, or straight ahead. And so it went for hundreds of miles and probably would have continued for hundreds more were it not for a fortunate break.

On May 6, 2002, three days after the start of the mailbox bomb spree, and after receiving a concerning letter from his son, Helder's father contacted the FBI because he was suspicious his son might be involved. While investigators started the search for him, the bloodhound team flew to his last known residence. Helder had left a note with one of his bombs that he left on the outskirts of his own neighborhood.[19] The team pulled scent from it, and the dogs took them straight to his apartment complex. They had finally closed in on him.

After Helder's apprehension, we learned that he had set off his mailbox tour with the goal of making a smiley face across the entire United States with his bombs. One eye was placed in Nebraska and the other eye split between Iowa and Illinois. Colorado and Texas made up the beginning part of the smile. Numerous graphic depictions of the artwork exist online for the curious. Helder never completed the mouth for reasons unknown.

No one ever got to testify in the Helder case, because he was declared "incompetent to stand trial," which was not surprising given his goals. (His mental state may also be why he never completed his mission.) Regardless, the Helder case showed true promise for the bloodhound unit, and the FBI was keen to capitalize on the potential. Rex and I weren't sure we were ready for prime time in terms of what the hounds could do, but prime time came calling anyway.

[19] Unfortunately, I am not at liberty to discuss the contents of the note, as even though Helder confessed to the FBI of planting the bombs, he has yet to go to trial for these crimes.

From 1992 to 2003 a serial rapist and murderer was terrorizing Baton Rouge. In September 2002, the department had no solid leads, and after more than a decade of dead ends the police chief was desperate to find some direction for his investigation. Rex said he would talk to the team and see if they thought they could offer any real assistance. I was a bomb guy. I was an explosives scientist. At first glance, my involvement makes little sense. But I had also assisted designing research programs to test the bloodhounds, worked several explosive and non-explosive related searches with the teams, and unfortunately impressed Rex with displays of analytical thinking. As a reward for these seemingly unrelated career tangents, I ended up deployed with a team of bloodhounds with the hopes of tracking down a monster.[20]

The most recent murder had occurred on July 12, more than two months prior to when we jumped into the case. With each passing day, scent degrades, dissipates, and gets covered over with other scents, making it hard to establish a trail. We also had no idea if the murderer was local or if he lived in a distant town and only came to Baton Rouge to choose his next victim. If the latter was the case, we'd have a slim chance of being able to track him. To that end, the dogs could be helpful. If we took them out and got no trail anywhere, it would suggest that the murderer was not in the area and had not been for a while. On the other hand, if we took the dogs out and they actually started to trail, it was possible the murderer was often present in the city.

Before we could do anything, we needed the scent of the murderer. We settled on three pieces of the most recent victim's clothing, including her bra, which had been forcibly removed. Now we had to figure out how many scents were on the evidence.

[20] I realize that at this point in our journey together, you might be cocking your head thinking, "But you chase down monsters for a living." Ideological monsters at least have a logic the scientific mind can comprehend. This type of primal monster, by contrast, I find incomprehensible.

We knew the victim had been dead for two months. This meant that she had stopped leaving odor trails with her car through town. If the killer was still scouting out other victims, his odor should be stronger. But there was another snag: others had handled the evidence. It was collected by a police officer; and handled by evidence technicians, and a forensic examiner. Each one of those three people lived in the Baton Rouge area, and each left daily fresh scents through the town.

As a result, the police officer, evidence tech, and forensic examiner had to be at every scene where we presented the bloodhounds with odors. Here's why: After a hound is given the odor, it walks past all the other people whose odor might be on the evidence and in the vicinity of the search. Smelling them at the scene, the dog realizes it does not need to chase after them. Instead, it looks for a trail put down by someone who is not there, and it follows that person's odor.

The very first time we introduced odor to the dogs all three began to trail. This meant that an odor present on the evidence was also in the area, and someone was leaving a trail of it around town.

Were this an episode of *CSI*, the hounds would have led us straight to the perpetrator's car, where we would pick up some fingerprints and apprehend him in a back alley right before he committed his next heinous crime, saving the day by sundown. But this case didn't resolve in a day. It went on for days—many draining days.

Baton Rouge is hot as Hell's sauna in September. So, we limited our trailing until early evening to allow rush-hour traffic to die down and to allow the heat to dissipate from ungodly to almost bearable, with temps in the low nineties and humidity in about the same vicinity.

Our whole team included three handlers, three massive bloodhounds, multiple police officers in uniform, Rex, an FBI technician, and myself. We rolled into locations in our big black SUVs. As we streamed out of the vehicles, like clowns out of a tiny car, the bystanders gawked. A six-motorcycle police motorcade surrounded us with lights flashing. With no further fanfare, someone put a dog on a harness, presented it with odor, and waited for what would happen. In a couple of places, the dogs just sat, immobile, but in most cases the dog's nose hit the ground, and it took off.

Bloodhound work, as it turns out, is not for the unfit. Dogs do not amble down the road; they run. If you want to follow them, you run, as well. I've been a runner most of my life; at the time of this case, I could run five to six miles without much trouble.

So, run we did. Down the major highways in Baton Rouge, we ran with the dog in the lead, handler in close proximity, and at least six other team members jogging behind. At least two motorcycles rocketed ahead to shut down all intersections in front of us, two motorcycles flanked us, and two bikes stayed behind to block traffic from running us over from the rear. Night after night, the dogs pulled ahead through the heat, sirens, and flashing lights, and we ran behind them, our bodies drenched with sweat. With thick air dragging us down, the dogs drove us ever forward.

For four nights, we ended up running an almost identical route through the city. And each night, no matter where we started, the dogs always led us to the outskirts of Louisiana State University, where the dogs wouldn't come to a halt, but would endlessly circle like water spiraling down a drain (as, increasingly, did our will to go on). The more we circled the same routes, the more I felt tangled up, like a dog who snares his leash around a tree and can't work his way back out.

Nothing made sense. We were getting nowhere. But underneath the frustration was the nagging sensation that the murderer was still in the area and he, too, was circling and meandering as he scouted out his next victim.

The last night of our investigation coincided with a memorial service and vigil being held to honor Gina Wilson Green, a forty-one-year-old nurse and office manager who had been raped and strangled in her home near Louisiana State University one year earlier. Our team decided to do our last run that evening from the scene of the memorial. The old cliché is that criminals always return to the scene of the crime. Serial killers have been known to take mementos from victims; it was possible the killer would drive by the service to relive the crime.

It was dark by the time the vigil was over. We took up positions in an area we felt offered easy accessibility to someone who wanted to drive by and observe the event. We presented odor to one of the dogs and off it ran. By this time, any of us could have run the basic route blindfolded. Just about every trail we took ended up going to the same highway, and eventually winding us up through the residential neighborhoods surrounding LSU. We hit the major road we always did. We hit the intersection, which always took us back to LSU. But this time the dogs kept going. We were on a new trail.

The advantage of having three dogs is that you have three independent detectors. With one dog taking us in a totally new direction, we let the team run a bit more before bringing out another to double check our result. The second dog also took the new turn off the highway. Eventually, we ended up in the parking lot of a Walmart—an area the dogs had never brought us before. Out came the third dog, who also took us into the Walmart parking lot.

We told the team to get the surveillance tape from the Walmart right away to see if anything stood out. But without knowing what the suspect looked like, and short of the perpetrator walking out with a shovel, trash bags, rope, and a copy of *Body Disposal for Dummies*, this was another long shot. To this day, though the killer was eventually caught, I still don't know why we ended up in that parking lot.[21]

With one too many unexplained events behind us, and four sixteen-hour days starting to drag us, one of the handlers mercifully suggested, "Let's get some steak and a couple of beers." Nothing sounded better at that moment. But before I could exhale, a police cruiser rolled up with lights flashing and one of our partner Task Force officers jumped out of the car declaring, "We need your help." All of us froze in our places. I can still hear his next words: "We just had someone shot dead outside a local business. This never happens in Baton Rouge."

Mrs. Hong Im Ballenger was leaving the Beauty Depot, a supply house she managed, at about 6:30 PM when a man confronted her by her car. She was shot once in the head and died almost instantly. A witness later told police the shooter grabbed her purse and ran toward a small park nearby.

We loaded up the canines, and police cars escorted us to the shooting scene. We arrived to find a semi-empty parking lot with a body lying next to the curb under a tarp.

What could we do? The killer had been in the vicinity only an hour or so before. The trail was fresh, but how could we get his scent? We fanned out and started to look for a shell casing. A sharp-eyed member of the team spied a side-view mirror on a car parked in the lot that was broken. In a remarkable break, we pried apart the side-view mirror and found a bullet fragment. The same bullet that had traveled through the victim's head split into pieces, lost energy, and exited to lodge itself in the side view mirror of a car. We had a potential source of scent.

---

[21] The killer we were chasing was Derrick Todd Lee. Lee began his life as a serial killer in 1998. In the end, he was linked to eight known attacks (with only one survivor, whose attack was luckily interrupted by her son) and suspected of two others. His last-known attack was in 2003. This is yet another example of how popular crime dramas do reality injustice. TV wraps up cases in sixty minutes. Reality can take years of close calls and frustrating dead ends.

In the movies the cops push a T-shirt under the dog's nose to give it scent, and off the baying hound runs. This is not the way it works in the real world. Scent is pulled from an item of evidence using various techniques. The FBI used a device called a Scent Transfer Unit (STU). For lack of a better analogy, it is akin to an expensive vacuum cleaner—a device the size of a handheld dust buster. The STU functions in an almost identical manner, except there is no vacuum-cleaner bag. The STU holds a sterile square of cotton gauze. You put the evidence on a screen covered by the gauze and turn on the STU. With large evidence items, you move the STU around the item so that the gauze comes in contact with it. Once the scent is pulled, you either present it to the dogs on the spot or store it in the freezer inside sterile mason jars until it's needed.

In this case, one of the team members and I took the bullet and collected odor with the STU. Rex readied one of the bloodhound teams. We had no way of knowing if enough scent was present on the bullet fragments for the dogs to take scent, but it was all we had.

The handler presented the gauze to one of the dogs, and it took off with more energy than I had ever seen. The trail was hot, and the dog was energized and determined. Everyone pulled their guns and the chase ensued. Rex called out to one of the uniformed officers to grab a car and bring me along behind. Again, I am a scientist, I am not a field agent, and, as such, I do not carry a gun. I was also not issued body armor at that moment. Not for the first time, or the last, I was the only one coming to the party with neither weapon nor protection. So, I jumped in the back of a cruiser and followed the chase. The luxury of riding in the back of an air-conditioned police car, and not running in the oppressive Baton Rouge heat, was a hardship I was happy to shoulder.

Within a mile we ended up entering a run-down trailer park. The dog slowed and its head dropped. I could feel the tension in the team. The uniformed officers had no idea what this change in body posture meant, but we did.

When the scent is strong and the trail obvious, bloodhounds don't have to work extra hard. It is like coming home and smelling something bad. It does not take much effort for a person to quickly figure out which room the odor comes from. However, once in the room it takes some sleuthing to figure out the exact source of the odor. Bloodhounds work the same way. The minute the dog's nose dropped down, it meant that there was a large volume of scent in this trailer park. The shooter probably came in and out of here on a regular basis, leaving his scent all around. With all this scent scattered about, the dog was working hard to determine the freshest trail.

I rolled down the window and could hear the dog breathing in and out quickly, snorting out big puffs of air, which indicated that the dog was in a pool of scent.[22] Rex let the police know we were close. We had about six officers with service pistols drawn and three police cars with lights flashing all creeping down the street behind the hound. Eventually, it stopped near one trailer. At this point, the bloodhounds were pulled out of potential harm's way.

Rex, weapon drawn, stood off to one side of the trailer's door. Another police officer, also with his gun drawn, stood to the other side. Other officers flanked the windows, their guns drawn. At that moment, the policewoman driving our car pulled right in front of the trailer, with only about twenty feet between where I sat and the trailer's front door. I was directly in the line of fire, should a gunfight happen. Certainly, I would take the first bullet. Heck, I might take all the bullets. Somehow, I managed to retain my civility and leaned forward to politely ask her if she could

[22] Scent pools are areas saturated with the scent of a person. The pool is deepest in the couch where you sit daily, soaks your whole house, and gets shallower the further you travel from home.

move up a little bit, as I really did not need to observe the ensuing firefight in Sensurround.

The police knocked on the door—the kind of authoritative pounding knocks one can't miss—multiple times. Nothing. No response. It appeared no one was home. The tension in the air dimmed, before snapping back like a high-tension wire crackling in the rain. The door of the trailer next door flung open. A young, shirtless African American man sauntered out, shirt in hand. Staring down a gaggle of officers with guns drawn amid multiple lit-up police cruisers, he nonchalantly draped his T-shirt over his neck and ambled towards a trailer behind the one from which he'd just emerged.

In the end, the dogs had taken us to the doorstep of something. Of that we were certain. But what? We couldn't be 100 percent sure. We advised the officers to conduct a thorough canvass of the trailer park to see if they could roust up any leads. Then we gathered our teams and ended our watch, too tired for anything but an early night and some exhausted sleep. On the plane ride home, I came to the conclusion that I needed to get back to focusing on bombings. I really did not need the extra stress of the bloodhounds.

I had been home a week, when on Wednesday, October 2, 2002, a bullet slammed through the window of a Michaels hobby store in Aspen Hill, Maryland, just missing a cashier. An hour later, a second shot took the life of a National Oceanic and Atmospheric Administration analyst in the parking lot of a big-box retail store in a neighboring town. This marked the beginning of a three-week grip of unease, paranoia, and fear, which settled across the National Capital Region. This time, I was not launching to some far corner of the globe to investigate the handiwork of the unconscionable. This time, the monsters had come into my backyard.

The next day, things spiraled into pure chaos quickly. Four people were executed within hours, and a fifth was killed that same evening. This was all too close to home, and though I'd wanted to wash my hands of bloodhound work, I also didn't want blood on my hands. This seemed the wrong time to stick to my conviction of backing away. Upon arriving at work that Friday, I went straight into Rex's office and asked him if he thought the bloodhounds could help resolve this situation. Rex told me that the bloodhound teams were actually due to come into town the next week on a separate mission. He also paused for a second before saying, "We can't do everything." It was hard to hear. I was new enough to deployments that I felt a drive to save the world, but I trusted his investigative instinct. As it turned out, another victim was wounded that same afternoon.

Over the weekend, things quieted. I took Rex's insight as sound and calmed down somewhat, but everyone in DC felt the pressure. At the time, my two sons were six and three. Their schools stopped allowing kids to go out and play for recess. There is no way to explain to a child the horrific reality of random murder, especially when you can't understand it yourself.

At 8:09 Monday morning (October 7), while driving to work, I received a breaking-news gut punch. Thirteen-year-old Iran Brown had been shot in the abdomen and critically wounded as he arrived at Benjamin Tasker Middle School. I beelined to Rex's office. Upon seeing me standing there, he just looked down, took a deep breath, and said, "I know. We are going to run the dogs."

In Louisiana, we chased a ghost. The odor trail was old; the killer could have been anywhere within a hundred miles. But in the sniper case, we were chasing a suspect with a high-powered rifle and proven ability and desire to kill.

Getting odor was the first task. At the shooting of the boy, the snipers left behind a shell casing and a Tarot card (the Death card) near the middle school where the attack occurred. We were in luck. We pulled odor from both pieces of evidence. With murders taking place in Maryland, DC, and Virginia, a huge geographical area was open for search. There was no way of knowing if the killer was a resident, or someone just randomly driving around with no set home location. By the time we started running the dogs, the snipers had shot eight people. We had eight potential locations to start from. The cluster of locations in Maryland from the first day's shooting seemed the best place to begin.

During this time period, I worked two shifts. By day, I was in the lab conducting training as a bombing investigator. In the evening, I would meet up with Rex and the bloodhounds to visit locations and run trails. I would get home at one or two in the morning and start the whole sequence over again after just a few hours of sleep. It was like a recurring nightmare, running through one suburban strip mall after another. Getting nowhere. We were in and out of the command center, getting investigative insight from the police working the case, and then heading back out to another site. Once again, we were going in circles.

On Wednesday, October 9, at 8:18 PM, a man was shot dead while pumping gas in Prince William County, Virginia. We were running a trail at the time when Rex got the call. The DC beltway is famous for traffic, and it was a miserable night for it. But the murder had just happened, and we were mere miles away. We had a chance at a fresh trail. With sirens blaring, we sped down the shoulder of the beltway between traffic and the barriers. One car was a little too close, and we knocked off the side-view mirror as we sped past.

Once on location, the chase was on...until it wasn't. From the crime scene we ended up in a cul-de-sac with every house around us dead quiet. We had not been able to make any clear trails. We stood at the top of the hill looking down into the dead end in utter silence. In the sparse quiet of the night, it dawned on me I could have a rifle trained on me right then. Once again, I was the only person on the scene without weapon or body armor. Unfortunately, the trail stopped there, and we called it a night.

We had gotten nowhere with the bloodhounds, and the team decided to stand down for a couple days, so I had the weekend off. We attended a friend's wedding in New Jersey, but the weekend fun was short-lived. Rex called me on our way back home to tell me we had a fresh lead to follow. I never made it home; my wife dropped me off outside the crime scene, and the madness started again.

On Monday, October 14, at 9:15 PM, Linda Franklin, an FBI intelligence analyst, was shot dead in a covered parking lot at Home Depot in Fairfax County, Virginia. One of our own had been taken.

Sometimes this FBI life has me pulling double duty; I had experienced that for months with dual position as "lab-scientist-by-day, bloodhound-runner-by-night." But sometimes one part of the job has to take precedence over the other.

Around the time Rex, the bloodhound team, and I were mourning the loss of one of our colleagues in the active DC sniper case, my other life—that bomb-expert life—called and wanted me back.

In a big way.

I would be bound for Bali, where, as you'll recall, I would be faced with the biggest bomb blast in an urban area I had ever seen.

After a few long days in Bali, analyzing shrapnel, estimating charge placement with sticks and strings, and scouring scarred buildings, I collapsed in my hotel room and clicked on the CNN international newsfeed, mostly for the company of background noise and distraction. I was staring at the screen, zoned out, when a man—Lee Boyd Malvo, who we now know was one of the DC snipers—walked handcuffed out of a side door at what appeared to be a police station. Staring at the man on my TV screen, the floor seemed to drop out from underneath me and I was transported to the backseat of the police cruiser in the Baton Rouge trailer park, watching the young man walk out of the neighboring residence. I couldn't be 100 percent certain, but somewhere inside me was the sinking, unshakable feeling that the young man being arrested was the same one who had so unnervingly, nonchalantly sauntered out of the trailer in Baton Rouge weeks ago.

Back in the States after the Bali case ended, the flashbacks hit me again. I waited weeks for them to pass. They didn't. Unable to shake the terrible feeling, I went into Rex's office and shut the door and told him I needed to confide in him. As I launched into my suspicions, Rex stopped me and said, "Let me guess, you keep thinking of Baton Rouge?" I was floored and relieved. Rex was having the same experience.

Our memories—and gut instincts—might have been in synch and correct.

The murder we worked in Baton Rouge on our last night was actually committed by the Beltway Snipers. It was their last murder in a spree before they started killings in the DC area.

I wish reality was truer to fiction and the good guys had stopped the bad guys on that fateful night in Baton Rouge. In a world where the dogs brought us definitively to Malvo, we could have brought him into custody. But bloodhounds are not a methodology that results in universal probable cause, as the law says. We told the local cops that the trail ended in this area; we found a neighborhood with a heavy scent signature. Law enforcement then needed to take this info and walk the neighborhood, talk to people, and develop a search warrant. If it was Malvo, he could not be ID'd by the dogs. There was nothing that could be done at that time.

We would always wonder if instinct was right.

As a doctorate of science and not a fictional Dr. Doolittle, I also had to come to the realization that scent is a complex matrix. We have no idea what makes it up. How do bloodhounds differentiate odors between, say, twins? Does it go all the way to the DNA of a person? There are too many questions without answers. We simply do not know. Without this most basic of knowledge, any attempt to optimize the technique hits a hard wall.

Rex made it the last part of his life's work to try to understand. He spent years running trials and making great headway in forwarding the science of scent detection. But he died before seeing those paths forward through to meaningful fruition. He was passionate about everything he did; his death left a void in the lives of everyone who worked with him.

As for the bloodhounds, the FBI eventually stepped away from them. And I stepped into another mystery, the likes of which I'd never seen.

CHAPTER 9

The Collar Bomb

On Thursday August 28, 2003, at approximately 2:30 PM, Brian Wells walked into the PNC Bank in Erie, Pennsylvania, with a homemade shotgun in one hand and a ticking time bomb strapped around his neck. Within an hour, the bomb would explode, killing Wells instantaneously and launching a seven-year investigation. The full story of the Collar Bomb case, starting with events leading up to the failed bank robbery and ending with the conviction of Marjorie Diehl-Armstrong in November of 2010, is complex and at times unbelievable. For those interested, this saga has been documented by retired FBI special agent Jerry Clark in his book *Pizza Bomber: The Untold Story of America's Most Shocking Bank Robbery*.

Almost as convoluted as the saga behind the field investigation is the inner workings of the bomb itself and the forensics that went into its analysis. Jerry Clark was the special agent who worked the field investigation for almost a third of his career in the FBI. Over the course of this time span, I had the pleasure of many interactions with Jerry. I was the lead forensic examiner who managed all the forensic aspects of the Collar Bomb case. To this day, it is the most complex and bizarre explosive device I have ever encountered.

The Collar Bomb case is unique in many aspects. From the outset, it was striking enough that the FBI gave it the designation of "Major Case 203," meaning it was the 203rd major case in FBI history. Typically, such a label is reserved for massive attacks such as the Oklahoma City Bombing, or those witnessed on September 11. With the designation of "Major Case" comes an obligatory code name to cement its place in the FBI lexicon. The Oklahoma City attack was called OKBOMB, the serial bombings of Ted Kaczynski were dubbed UNABOM, and the case revolving around the attack in Centennial Park in the 1996 Summer Olympics was CENTBOM. This case became known as COLLARBOMB.

As a forensic bombing investigator, my approach is somewhat different from other forensic disciplines. A fingerprint examiner, for example, does not need to know much about the crime scene to determine if a print of value is contained on a piece of evidence, or that a print found on evidence can be linked back to a particular suspect's known prints. However, to properly evaluate how a bomb functions, I need to know as much as I can about the scene of the bombing. Bombs can be initiated via a wide array of measures.

In the Collar Bomb case, the bomb could have been set to go off at a certain time, it could have been triggered by some action of Brian Wells, or it could have been remotely set off by someone from halfway across the world if the proper electronics were contained within it. The actions taken by Brian Wells during the last tragic hour of his life became very important to me even before I saw the first item of evidence come through the lab's doors.

When Wells walked into the PNC Bank that fateful day, he was wearing a Guess T-shirt stretched over the bomb, which was pressed against his chest and held in place by a collar ratcheted around his neck. Sporting a homemade shotgun made to look like a cane, he made his way to a line, grabbing a lollipop as he did so. A teller, thinking he was saddled with some type of medical device (and feeling concerned for him), called him forward.

Wells's behavior entering the bank gave everyone pause. I know if someone had strapped a bomb to me and told me to rob a bank I would not proceed to stand patiently in line. In all likelihood, I would also not pause to grab a piece of candy. It struck me as the behavior of someone who either had control of the bomb, did not believe it was real, or knew something that reduced his anxiety about its presence on his person.

Wells handed the teller a four-page note that explained how he was a "bomb hostage" and demanded money. By this point in my career, I had seen a fair number of bank robberies. It may seem odd for a bombing specialist to work bank robberies, but here's little-known fact: a good number of bank robbers use hoax bombs as part of their robbery scheme. They believe leaving behind something that looks like a bomb will slow down police response and buy extra time. New examiners in the Explosives Unit tend to work a substantial number of hoax bombs in the early stages of their training to learn the inner workings of evidence flow and report writing.

The note that Wells handed the teller had a section for her, a section written for the bank manager, and a long litany of threats aimed at bank employees and police if anyone tried to stop the robbery. It even had two options outlined for the folks being robbed. The bank could either hand over $250,000 and no one would get hurt, or, option two, hand over less money than that and risk harm coming to its customers and staff. It was obvious the people who hatched this plot had exposed themselves to too many hours of television crime dramas.

Faced with a man claiming to have a bomb around his neck, and a demand for money, there is no way that any bank employee takes the time to read through a rambling four-page letter of instructions and demands. Bank tellers always do the same thing regardless of what the robber does or if the bomb is real or a hoax; they hand over the money from their drawer and call the police at the first available opportunity.

Wells could have walked into the bank with a potato wrapped in aluminum foil and wires sticking out of it (also not an original creation, I'm afraid) and fared equally well. He walked out of the bank about ten minutes after walking in with approximately $8,700 for his efforts and both option one and option two cruelly snubbed.

If Wells's behavior in the bank seemed peculiar, it only became more so after he departed with his money. Instead of fleeing the scene of the crime, as anyone would be expected to, he remained in the nearby vicinity. Ten minutes after walking out of the bank he was arrested in the parking lot of the nearby Eyeglass World. After being arrested and handcuffed, Wells notified police he had a bomb on his person. An officer lifted up Wells's T-shirt and saw a box with a window in it, exposing, amongst other things, a digital clock, wiring, and what appeared to be a phone. The description of the bomb provided by that officer was the only insight I ever had into the device's construction, as I never got to lay eyes on it myself before it was disseminated into smaller pieces.

Wells was told to sit on the ground, the scene was evacuated, and the bomb squad was called. During this time, Wells told police he had been abducted and forced to wear the bomb with instructions to rob a bank. Wells was found with a four-page note addressed to the "bank hostage" on his person, which consisted of a set of instructions on places to go and items to search for that would help him remove the bomb. He was still at the first stage in these instructions when he was apprehended and arrested.

Once it was established that Wells very well might have a live bomb on his person, officers shielded themselves behind a patrol car and continued communication from a safer distance. Eventually, the press arrived at the scene and began filming this unprecedented standoff. Wells sat for nearly half an hour when police witnesses report they started to hear a beeping sound. The beeping went on for several seconds and then the bomb went off. From the time Wells walked into the bank until the bomb exploded, roughly fifty minutes had elapsed. Within a few minutes of the bomb exploding, the bomb squad arrived on scene.

For a small-city bomb squad, the Erie Police Department made good time for the callout. The fact that they arrived within minutes of the blast was most likely a blessing for them. Bomb techs have a variety of safety protocols, including wearing elaborate safety gear before approaching a suspected IED. However, when human life is in imminent danger, these protocols can be suspended at the discretion of a bomb technician. Had the techs arrived prior to the bomb going off, one of them might have approached Wells immediately without safety gear. Depending on the circumstances, this could have cost the life of the responding bomb tech.

As a forensic examiner, two critical pieces of evidence came to me from these witness reports. The first was the statement from the police officer who raised Wells's shirt and caught sight of the bomb. His description was not detailed, but it gave enough insight to shed light on a couple of pieces of the puzzle. The second was the beeping sound. As will be discussed later, the bomb did incorporate a digital timer. The question I was confronted with was whether or not the timer could have set off the bomb. I surmised it was not likely to have done so.

In most cases, when timers are utilized to set off a bomb, the bomb builder wires the device in a fashion that kills any beeping sound. It would be possible for the bomber to wire in elsewhere in the timer and keep the speakers active, but the electrical current needed to set off the bomb would have been released the second the timer went off. The first beep should have been the last thing heard. The timer was not the trigger; instead, it heralded the imminent closing of some other switch secreted within the iron box strapped to Wells's chest. It would take me months more to figure out what that switch was, and how it ultimately set off the sequence of events that killed its victim.

I would like to say that what followed Wells's death was a lock step, textbook example of law enforcement cooperation and unified operations. That does not do justice to the scene. Remember those clichéd scenes in crime movies where the FBI comes in and "establishes jurisdiction"? At the FBI, the old saying is you can't spell jurisdiction without "dic." Some look at the Wells case as a murder. That puts any investigation squarely in the hands of the Erie Police Department. With no federal nexus, a local police department may reach out for assistance to either the FBI or the ATF to provide outside expertise on explosives, particularly if the police had none in house. Often, feds show up at what appears to be a major scene to assist proactively. With three agencies on scene, someone needs to take charge. In COLLARBOMB there was some rigorous debate surrounding who would have primacy. In the end, one clear thing tipped the scales. The FBI has primary jurisdiction at the federal level over any bank robbery. It was decided the Bureau had the lead.

Once jurisdiction was established, the Wells case went forward in a fairly clockwork fashion. The state police, Erie police, FBI, and ATF all worked together in an investigation task force.

The FBI Lab took possession of all evidence for forensic examination. The FBI's Evidence Response Team (ERT) was called in to photograph, document, and collect all evidence.

In the Collar Bomb case the ERT was assisted by FBI special agent bomb techs. Bombs can be made out of literally anything. On-site decisions about what might have been part of the bomb and what might be just random trash lying on the street are challenging. Traditionally in bombing scenes, the caution-minded ERT adage was, "Collect it all and let the lab sort it out." By bringing seasoned bomb techs to assist in evidence recovery, the collection of meaningless debris is minimized, and lab analysis is much more efficient.

As far as evidence collection goes, bomb scenes can be tricky. A shooting scene is fairly basic. An ERT team looks for shell casings and notes bullet strikes. In most non-bomb-related cases, very meticulous work needs to be done to document where every single piece of evidence is recovered. But a bombing is different. A bomb is chaos made incarnate. Sometimes bombs can contain hundreds of pieces of fragmentation, which, for the most part, fly outward in a random pattern. It really does not matter where most of the fragments of an explosive device are recovered from. To me, what matters is how all the pieces of the bomb fit back together.

To compensate for the randomness of a bombing event, a typical scene is broken into zones. In the Collar Bomb case, Wells was sitting up when the bomb went off. This meant most of the evidence was thrown outwards from his chest. Only items that ricocheted off nearby solid objects were found behind him. Wells's body was one zone for evidence collection. The parking lot was another zone. A grassy field nearby, where some bomb fragments were thrown, comprised a third zone.

Items recovered from those zones were packed together. We typically teach a "pack like with like" strategy. Within each zone, we recommend that all pieces of wire be packed together in one bag, all pieces of battery be packed in another bag, and so on. This limits the number of bags and the administrative workload required to keep track of evidence. But in some cases, the ERT teams cannot shake old habits and treat bombings like shootings—resulting in an abundance of extra tracking and paperwork. Forests weep for the loss of their brethren when evidence like that rolls in the door.

I have always been a strong believer in the power of serendipity in shaping the course of a life. It seems to have had more effect than any hard-wrought plans in mine. Evidence was set to arrive at the FBI Laboratory late Friday, August 29. Monday was Labor Day. The boss came looking for someone to assign the case to and found only two qualified examiners still in the lab. Having just come back from a long holiday with family, I had no plans for the upcoming Labor Day holiday. My friend (Mark from the Bali bombing), on the other hand, was getting ready to sell his house and had just told me of his plans to spend the weekend painting it to get it ready for market. I told the boss I would take the case. In my mind I was thinking something along the lines of, "How hard can this be?" In the words of Bugs Bunny, "What a maroon!"

Most bombs in the United States are not very sophisticated. Fuzing systems, for the most part, are fairly rudimentary, and, as a result, the forensics—while often tedious—are manageable. Without realizing it, I had consigned myself to years of one of the most challenging and bizarre cases of my career.

I gathered up a team of lab techs to prepare for the onslaught of evidence. From experience, I knew that the minute evidence leaves the scene, investigators start wondering why the lab is taking so long to get results back to them. My crew had to work around the clock.

Forensic-based shows like *CSI* or *NCIS* present a fairly warped view of how my job works. I do not sit in some fancy painted lab surrounded by mood lighting staring intently into a microscope, only taking time to acknowledge others with some haughty retort. The only thing I share in common with my television counterparts is my general aesthetic appeal.

In TV shows, the examiner gets a few pieces of very poignant and obviously critical evidence to brood over as he saunters about in clean-pressed lab coats. In reality, bombs scatter small pieces of debris to the wind. Examiners get a truck (or plane) load of mangled, bloodstained crap to sort through. I typically wade through bags of absolutely indecipherable carnage. As for my lab coat, when I wear one, I can guarantee it is neither clean nor pressed.

When *Popular Mechanics* came to my lab to do a photo shoot for a profile article on me, they asked if I had a lab coat. I found one of mine crumpled in a corner and put it on. The photographer gave me a look of thinly disguised dismay and asked if I had a coat that was ironed; I quickly learned that his photo team traveled with a portable steamer. That may be the only time my coat ever saw a day without wrinkles. Had I known I'd have access to a steamer, I would have brought a couple suit jackets in for steaming, as well.

In the FBI Lab, when the evidence arrives, it is opened, documented, and parceled out by a team of young technicians who work under an examiner. Erie, Pennsylvania, was close enough to our lab in Quantico (Virginia) to have evidence driven down by the truckload. About 120 bags of packaged specimens were offloaded. Some bags contained a single piece of mangled plastic, while others contained dozens of small pieces of fiberboard, thin metal, and materials I could not even begin to identify.

Checking evidence in to the lab is not glamorous. It is painstakingly tedious work. One person sits at a computer and does nothing but document items as bags are opened to ensure everything communicated from the field is indeed in the boxes. Field evidence technicians collecting debris at a crime scene live with evidence for a couple days at most. In the lab, we have to live with it for years. Every piece delivered from the field has to have a flawless chain of custody that can be tracked back to the moment in time it was picked up from the scene. That means a great deal of paperwork, inventorying, and cross-checking. It also means frequent cursing and calls back to the field to sort out even the smallest administrative discrepancies.

I don't recall ever seeing that episode of *NCIS* where Abby sifts through sheets of paper, two computer screens, and countless layers of boxes and bags engaging in a perplexing conversation with a rookie field agent trying to figure out why someone felt the need to send in a cassette tape from a storm gutter (which happened to me during one of thousands of 9/11 evidence submissions). All this is done while trying to determine what is trash and what is evidence. Add to that an all-night marathon session fueled by pizza and stale donuts where lab techs parse through blood-soaked fragments of indecipherable debris, and a better depiction of true forensics begins to emerge.

Evidence check-in marks the start of a tremendously anal-retentive process. While one person inventories, two other technicians take photographs of everything. We photograph every item of evidence that comes into the lab regardless of how mundane it might seem at first glance, especially because the evidence goes through many steps before it comes to me.

Before I can look at it, evidence goes to three or four different examiners to be slowly disassembled and probed for valuable forensic clues. For example, bombers love duct tape. Duct tape is one of the few human inventions that actually holds back the forces of entropy. When evidence comes in covered in tape, there is no way of telling what is buried in the wrapped adhesive. And I can't open it up and peek. There could be a treasure trove of hairs and fibers, latent fingerprints, and DNA caught within the folds of the tape. When bombing examiners think that tape may be covering some crucial design aspect of the bomb, communications go out to the other examiners to take pictures at each step of their unraveling. In some cases, I work with other examiners as they take apart items submitted to the lab. There is no way to know what item holds the most crucial clues. Therefore, we photograph everything, and often we photograph the same items at multiple steps in their travel through the lab.

In the Collar Bomb case, I served two major roles. First and foremost, I was the examiner whose job it was to determine what the bomb was comprised of and how it functioned. That alone took many months. Of more immediate relevance, I was the examiner who developed the forensic plan of attack. For each of the hundreds of items that came into the lab, I decided which disciplines needed to examine them and the order in which the items had to go through those disciplines. The purpose of this evaluation is to put in effect a systematic flow of evidence from examiner to examiner and to ensure that no exam is conducted that destroys the potential for other disciplines.

In some cases, such decisions are easy, as certain items have established forensic protocols in place. Give me a piece of tape, for example, and it will first go to Latent Prints to have an external examination done under laser light to see if any obvious prints are on the outside. If nothing is there, the tape next goes to Trace Evidence to be unwrapped and searched for any valuable hairs and fibers. Trace knows how to preserve the tape they take apart so that the evidence can go back to Latent Prints a second time for an examiner to look at the full length of tape on both sides. Finally, the tape goes to our Chemistry Unit to be analyzed and to gather chemical information that is then used to compare against any tapes seized during follow-up searches.

In some cases, there is no obvious order of examination. In the Collar Bomb case, multiple pieces of angle iron, which were riveted and bolted together, were initially sent to our Latent Print Unit. Once they were exploited for latents, there were no further exams that seemed necessary at the time. However, any evidence can gain additional relevance as a case progresses. At one point, searches in the case started turning up metal plates and angle iron similar to the material that was used to construct the bomb. At that point, the original metal from the bomb was brought back for further chemical analysis by a metallurgist to compare against the items recovered from searches associated with suspects.

In a typical bombing case, I watch as all the evidence is unpackaged and photographed and fairly rapidly have a decent idea how the bomb was made. As I watched more and more items come through the lab in this case, however, I was stunned. Some items were immediately obvious: lengths of pipe from pipe bombs, wind-up kitchen timers, four AA batteries, multiple lengths of wire. But these easily understood items were soon dwarfed by an ever-growing list of baffling debris.

The first round of evidence that arrived in the lab was from the scene of the bomb blast and from Brian Wells's vehicle. The one thing that I noted by the time we had plowed through hours of evidence check-in and reached the end of the first delivery was that I had no collar. Admittedly, in the early stages of evidence check-in, things can be absolutely hectic and small things can go unnoticed. However, a massive steel collar is not one of those things. I called the field to try to ascertain what happened to this crucial component of the bomb. Terse communications informed me it was still attached to the body of the deceased and discussions on how to separate the two were ongoing. I knew enough to let it lie.

I eventually did get the collar the next day. It was hand delivered by the FBI bomb tech from Pittsburgh who deployed up to Erie to assist in the evidence collection. The bomb tech was in a talkative mood when he delivered the collar and felt compelled to explain to me just how he managed to take possession of it. The reason the collar created such consternation is that it had a series of wires wrapped around an odd plastic tube intertwined around the whole piece. The wires and tubing vanished into a box at the collar's base, which held the locking mechanism. The metal box was too thick for any field X-rays to really allow a good view into the inner workings of the collar.

In essence, no one could be sure the collar was not a separate explosive device. The safest way to remove the collar, according to the still-rattled bomb tech who witnessed it, was rather grisly. The bomb tech proceeded to explain to me in graphic detail how he had been in the morgue to evaluate the safety of the collar while the medical examiner proceeded to decapitate the victim in order to remove the device.

A more trained psychologist might have tried to dive into the tech's feelings during this time, but I just looked at the collar sitting in its plastic bag with obvious blood splatters adhering to the inside and wondered if I was holding a bomb. There are moments in my career where I can remember wondering, "How the hell did I get here?" Listening to the bomb tech's gruesome recounting of his day at the office while trying to figure out what to do with a blood-covered massive steel contrivance that looked like a handcuff shackle on steroids was one of those.

While the bomb tech went off to share his story with the next unsuspecting lab technician who asked how his day was going, I went off to examine the collar. I noted that it had one arm that swung freely into the lock box. This arm was stopped from going all the way through by a metal screw blocking its movement. I also noted I could not pull the arm out of the lock box. In the position the arm was resting in, I could only see partially into the metal box. It looked like the tubing and wires truncated into the box but were not attached to anything. Part of the box remained obscured to me.

The collar had to go to forensic analysis, but I knew I had to make sure it was safe before moving it through the lab. I could see that part of the locking mechanism consisted of two small padlocks, the size of typical luggage locks. The key slots for these locks were visible on the outside of the metal box. Nowhere in the evidence had any keys been submitted.

Before I joined the FBI, back when I was a young scientist making explosives in the New Mexico desert, I hired a lab tech to help with my research. An amateur locksmith, he sometimes brought out his lockpick and taught me how to pick locks. I found it so much fun, I bought my own set of picks, which I have to this day.

I was fairly certain I could pick the collar's locks. I was also fairly certain there was no explosive device secreted within its confines. I was uncertain enough of these certainties that I told everyone to leave the room, but to stay within hearing range. They implicitly understood why. Putting the collar on the table in front of me, I was able to pick the two locks in under a minute. It took me longer to figure out that once the locks were open that there was a release lever that had to be pulled to allow the large cuff to be opened. It was a relief to see the insides totally visible and be 100 percent certain the device was safe. We now had all the evidence from the scene in hand and ready for full forensic exploitation.

In forensic science, there is a phenomenon referred to as the "CSI Effect." Put simply, people watch way too much TV and movies. While this may keep the unruly masses docile (in the view of conspiracy theorists), what is, in most cases, meant as mindless entertainment slowly starts shaping a shared worldview. This would be fine if the view was not so distorted. Originally released in October 2000, the TV show *CSI* introduced viewers to the world of the forensic examiner. It did so in a manner that warped the forensics reality.

Those who don't work in the field find it hard to believe a simple TV show, and its three demonic spawn (*Miami*, *NY*, and *Cyber*), could really have such far-reaching effects. But *CSI* not only created spinoffs, it energized a genre that included other shows such as *NCIS* and *Bones*. Together, such shows painted a picture of the actual job of forensics more akin to fantasy than fact.

Interest generated by the romanticizing of the forensic field has actually created a shift in the college landscape. In 1997, three years before *CSI* emerged from the dark underworld, only three to five bachelor degree programs in forensic chemistry and forensic science existed. By 2007, over thirty forensic chemistry programs and almost fifty forensic science bachelor programs existed. A 2017 search of the College Board website for higher education institutes listing forensic science programs resulted in 218 returns.

In my world, where the CSI Effect is most profoundly felt is in the court of law. Jurors receive most of their exposure to forensic concepts from shows whose grasp on science is tenuous at best. Jurors expect prosecutors to come into court with DNA, fingerprints, and an occasional dramatic plot twist to keep them engaged. What they get rarely, if ever, meets these lofty aspirations. In the decade I actively worked criminal bombing cases as an examiner, I can think of maybe three cases where I had fingerprints of value, and none where I had DNA.

Forensics did not ride up like a knight in shining armor to rescue the Collar Bomb case, but it did help guide investigators down productive avenues of pursuit...eventually.

In the first weekend of the Collar Bomb case, we had evidence from the bomb itself, Brian Wells's car, his house, and a series of sites listed in the four pages of instructions directing him where to visit. These were the first items to be examined.

Before the advent of the FBI's current computerized evidence-tracking system, items were broken down into two broad categories: Q items (questioned), or K items (known). The delineation between Q and K was often a source of heated debate, but, at its most simple, Q items were derived from crime scenes, and K items from subsequent searches. Anything from the car, seat of the explosion, and bank robbery was denoted as Q. By the time all evidence from the scenes was delivered to the lab, we ended up with 214 Q specimens. Remember that a specimen could be a single screw or it could be a bag with dozens of small pieces of fragmentation. In the early stages of the case, our primary goal was to learn what forensic insight the Q items in our possession might provide.

In a case the size of COLLARBOMB, the FBI Laboratory goes into round-the-clock operations in order to exploit the evidence as quickly as possible. A victim has been killed and the bomb builder(s) assumed on the loose. Any insight that science can offer is crucial to assist investigators in focusing their energies. Although they didn't break the case open, two forensic disciplines provided field agents with information within the first twelve hours.

The most productive information came from a latent fingerprint examination of the four-page note that Wells was carrying with him following the bank robbery. This note gave detailed instructions of where Wells was to proceed the minute he got money in hand. Almost every print found on the pages belonged to Wells, but one print not attributable to Wells was found. Upon discovery of this print, examiners quickly contacted the field investigators to share the results. One astute Pennsylvania

Trooper told his fellow investigators to hold off on celebrations because he believed his partner was the individual who collected that evidence. He noted that his partner was always "finger fucking" evidence and we might have gotten one of his prints due to his carelessness. Unfortunately, a comparison with the prints from the digital fornicator did show him to be our source/culprit. Strike one against us.

It might seem useless to have a note covered with the fingerprints of the victim who had it in his possession. However, sometimes physical evidence gives credence to a story or narrative. In his last hour on Earth, Wells told investigators he was abducted while delivering a pizza, had a bomb strapped to him, and was told to rob a bank. He told the arresting officers that he was given a note to follow in order to free himself from the bomb.

There is no doubt that Wells looked at all the pages of the note. His prints were on all of them. Had the note been a prop, and Wells had no need to read through it, it might be expected very few of his prints would be found on the pages. The significance of the prints is open to interpretation, and Wells's level of involvement was never 100 percent pinned down. But his obvious heavy handling of the pages helped put together a picture of events as more evidence was collected.

Another moment of jubilation in the lab came the first evening when we got a report back from our Questioned Document Unit that indented writing was found on the note pages. Indented writing is defined as "the impression from the writing instrument captured on sheets of paper below the one that contains the original writing." Next time you are in a hotel room, check out the writing pad to see what indented traces remain. In what at first seemed like a remarkable score, our document examiner found not one but two telephone numbers indented on the note pages. To get any indented writing in a case is rare. To get something as potentially valuable as a phone number in the indented writing is even rarer. To get *two* numbers is unheard of. The phone numbers lacked area codes but had the rest of the seven digits clearly visible. The excitement spawned by this discovery energized us through the first long night.

It has been said that fate has a wicked sense of humor. Like just about everything else in the Collar Bomb case, the two phone numbers slowly turned from a beacon of hope into a sneering taunt. At first, investigators looked into local area codes for Erie to try to track down what the numbers referred to and how they could be associated with the bombing. Every avenue was met with a dead end. Eventually, an ever-growing radius of area codes were tried, and as the ripples expanded, hope in ever converting the numbers into investigative leads diminished.

During the 1950s, a grim set of experiments showed the power of hope. Rats were placed in a tub of water. The first group of rats were allowed to tread water until they gave up and drowned. Most rats gave up fairly quickly. A second set of rats were taken out of the water just before indications showed they were about to drown. The saved rats were then placed back into the water. The rats that had been rescued struggled to stay afloat drastically longer than those who had been left to their own devices. Thus, it was not physical limitations that caused the demise of the unaided rats, it was attributed to a lack of hope. In an oddly analogous fashion, forensic examiners never, ever give up on a positive forensic result. The discovery of the phone numbers was the hand that lifted the team out of the water early on in the case. We continued to tread water for years after.

Rat analogies aside, there have been plenty of high-profile cases where evidence does not reveal its value for years. One case is straight from the files of the FBI. Two pipe bombs sent through the mail in December of 1989 killed a federal judge and a prominent civil rights attorney. Two more similar bombs were interdicted shortly thereafter by police and disarmed by bomb technicians. The ensuing investigation comprised a major FBI case, known as VANPAC.

Forensic leads in the case were limited. The bomber, Walter Moody, had gone to extreme lengths to cover up his forensic tracks. However, a fingerprint was found on a letter sent along with one of the disarmed bombs. No match could be found. But a forensic clue doesn't have to be a flash of light; sometimes it is a smoldering ember. In the Moody case, a bomb technician recognized some features in the bomb as being similar to those he had seen seventeen years earlier. The similarities between the bombs brought Moody into the investigators' crosshairs.

However, even with Moody's prints to use for comparison, the investigators could find no match between the fingerprint discovered on the letter and this prime suspect. As the case progressed, Moody's ex-wife became a crucial witness. She explained to investigators that Moody had ordered her to disguise herself and drive off to distant locations to shop for items he needed to produce his bombs. In one case, she drove seven hours to a copy store in Florence, Kentucky, to make copies of documents she was told not to look at. While at the store, the copier ran out of paper. A helpful clerk refilled the machine for her.

In the case of an ex-wife testifying against ex-husband it can be difficult to ascertain motives and veracity of witness statements. However, in the Moody case, the fingerprint turned out to be that of the copy store clerk from the top page of the stack of papers he had loaded into the copy machine for Moody's wife. After months of frustration, the source of the print was unveiled and produced hard evidence to corroborate the story of a crucial prosecution witness.

As a bombing investigator, I knew full well the dramatic story of the Moody case, and I shared it with other examiners on the Collar Bomb case to give them perspective, that sometimes evidence takes time to reveal its true worth. Despite the fact that COLLARBOMB never had the same made-for-television moment as the identification of the VANPAC fingerprint, what our team lacked in leads we made up for in hope.

In the early days of a high-profile case, the relationship between the field investigators and the laboratory scientists is very dynamic. In some cases, investigators have ample leads and clues to develop and oftentimes arrest suspects quickly. When this happens, the lab basically looks to see if any of the forensic evidence corroborates investigative findings. Such circumstances create a much more relaxed environment.

At the other end of the spectrum, the field investigators have absolutely no investigative leads to follow. When this happens, field investigators transfer their hopes and anxieties to laboratory examiners. The Collar Bomb case was just such a situation. A man with a bomb strapped to his chest was blown up in broad daylight after robbing a bank. Investigators had very little to go on. In his last hour of life, Brian Wells provided precious little insight into his attackers. To rub salt into the wound, the two indented phone numbers the lab did find were turning out to be a source of frustration instead of jubilation.

The only avenue investigators had to gain insight was to talk to Wells's associates and to visit the site from which he was abducted. In terms of associates, his fellow pizza delivery drivers were first in line. Robert Pinetti worked side-by-side with Wells at Mama Mia's Pizzeria. During the early stages of the investigation, agents spoke to both the owner of the pizzeria and Pinetti. Investigators were too busy to conduct a detailed interview with Pinetti and had planned to return with further questions later that weekend. That interview would never take place.

Looking back with the full perspective of time, there were certain telltale signs that the Collar Bomb case would not follow any established norms. On Sunday August 31, three days after Wells's death by bombing, Pinetti was found dead in his home. Early analysis showed a potentially lethal combination of drugs in his system; Pinetti was a known drug user. News stories from the press noted that an FBI special agent in charge said there was "no reason to connect Pinetti's death to Wells's case." At the time, this was entirely accurate, but no one working the investigation had a good feeling about this turn of events.

I do not like coincidences. I fully realize that the universe is capable of producing unlikely events, and that a bunch of monkeys can knock off *Hamlet* (or at least an Adam Sandler movie script) if the stars align correctly. However, after a man is killed in as unique a fashion as Wells, and a close associate, who the FBI has a distinct interest in talking to, dies a couple days later, I just didn't buy the "move along, nothing to see here" story line. I had no idea how much more convoluted things were destined to turn.

Pinetti became known as Victim Number Two by the bomb folks who worked on the case. The story of Victim Number Three is rarely told and known by only a small number of us. Oddly enough, it all started with a toy.

One of the items of evidence in the Collar Bomb case was the remnants of a toy cell phone. The purpose of this item was not known at the time, as the device had yet to be reassembled, but it was possible to narrow down the manufacturer. This toy was obviously old, as it depicted an outdated phone style. Oddities like this can sometimes generate productive leads. Typically, when an odd item is identified as part of an IED, investigators attempt to determine where the item is sold to seek out potential purchasers. To find such facts, it's often most productive to talk to a manufacturer or distributer. Investigators were able to track down the phone's source as a toy company located in California. An FBI bomb tech from Los Angeles was tasked with the job of visiting the toy company and finding out as much about the item as he could.

I have come to learn that the presence of an FBI agent on a doorstep asking some questions is greeted in many different ways. In some cases, folks get immediately paranoid. In others, people go well beyond any reasonable call of duty to lend assistance. In the case of the toy cell phone, the reception was blatantly hostile. To protect the tech's identity, I'll refer to him here simply as Kevin.

Kevin met with the president of the toy company and his lawyers to learn more. Apparently, when Kevin attempted to gain information about the toy phone, the president of the company went into a defensive posture, shouting through a thick German accent, questioning how the FBI knew the toy came from his company and berating us for blaming his product in any way (which we were not doing in any fashion). The word Kevin used to describe the gentleman was "apoplectic."

After venting at Kevin, the president stormed out of the room. The lawyers representing the manufacturer apologized and said the company would offer any assistance it could to the FBI. This would have all ended well except for the fact that the president of the company, on top of being highly agitated, was also of advanced age. That night, he went home and promptly expired of a heart attack. Thus, Victim Number Three was added to the Collar Bomb roster.

Working in the world of law enforcement tends to breed a dark sense of humor. Mine was already fairly murky before I even started working for the FBI (a characteristic I would like to officially blame on my mother, since I have the soapbox). Upon hearing Kevin's story, I told him that the bad guys seemed to be doing a good enough job introducing fatalities to the scoreboard, and I really did not see the utility in the FBI adding to the body count. His response was a simple, timeless two-word retort I seem to engender a great deal in my daily interactions with colleagues.

Within two days of the evidence arriving in the lab, I sat with Rex Stockham, who on top of being a bomb tech was an excellent machinist, and worked out a full list of materials used to produce the collar bomb. Such a list is crucial to inform field investigators what they are searching for when they're interviewing suspects. It also develops leads into what stores to start canvassing for purchases. The list was extensive. To give an appreciation for the scope of its complexity, a partial list of items included: angle iron, steel flat bar, schedule 40 steel pipe at one-inch diameter, wire mesh, machine screws, miscellaneous screws (#6 and #10), rivets, brake line, fuel line, hex bolts, washers, metal all-thread rod, Masonite, fiber-reinforced composite board, batteries (AAA and AA), brake line holders, solder, wires (red, green, yellow, parallel zip), cotter pins, gray paint, Walmart digital timers, Sunbeam mechanical timers, GSM toy phone, silver duct tape, flexible plastic tubing, paper labels, miniature brass locks, numeric tumbler lock, wire nuts, crimp-on ring tongues, and springs. Every item listed had dimensions provided, as well.

During the early stages of the investigation the local FBI office had very few leads to go on. A combination of information provided from tip lines and interviews with Wells's associates resulted in some early suspects. One of these was the "business manager" of a young lady who was known to exchange sexual favors for drugs and financial gain. This entrepreneur worked at a factory that specialized in custom plastic parts. He became known as the "Primo Suspecto."

The one facet of Primo Suspecto that was most enticing to investigators was that multiple associates mentioned to the FBI that he had spoken to them in the past about how, if he would rob a bank, he would do so by strapping a bomb to his victim to extort money. In my experience, most people don't go around idly speculating on how they would commit violent felonies. Of the few who do, even fewer then feel compelled to share their thoughts with those around them. Most likely, Primo Suspecto was simply spouting off asinine diatribes like this to impress whatever particular cadre of lowlives he was sharing a crack pipe with at the time. However, this little bit of braggadocio garnished him a great deal of unpleasant attention.

It was not hard to convince a judge that Primo Suspecto's vocalized plans were a close enough match to the events in the case to justify a search warrant. Soon enough, we were searching his attic apartment and his employer's machine shop. I can still remember him lingering outside the apartment while we collected hardware from his tool chest to compare against tool marks on the evidence. All the while he loudly vocalized the injustice we brought to his life. His one-man play consisted of him repeatedly yelling to the assembled audience of spectators outside "Look at me! I'm with the FBI! I've come all the way from Virginia to fuck up your life!" In the end, he was found to have no connection to the Collar Bomb case and was ostensibly a victim of his own bad life choices.

While we were still analyzing all the search items coming into the lab, a critical break presented itself. Ironically at the time, not only did we not know it was a break, but also everyone involved in the Collar Bomb case—from the investigative side to the forensic examiners—initially tried to divorce the event from the case.

On the day of the robbery, an order was called in to the pizzeria where Wells worked. The order set up delivery to a remote site that housed a television transmission tower. It was simply referred to as the Tower Site by our team. A dirt road led to the Tower Site and dead-ended there. Adjacent to the road was a single residence belonging to a man named Bill Rothstein. I was told in the early stages of the investigation that police attempted to make contact with Rothstein to question him about whether he saw anything the day Wells had been abducted from the site but were having no luck communicating with him. Other more pressing leads kept investigators from pursuing interviews with Rothstein.

On September 20, several weeks after Wells's death by bombing, Rothstein called 911 to notify police that he had a freezer with a dead body in it on his property. I recall that day vividly because, at the time, we were still trying to gain any forensic clues we could from the evidence and subsequent searches. Frustration had set in, as no productive insights were emerging, and the indented numbers were still eluding explanation. At about eight in the morning that day, I got a call from the local FBI bomb tech—the same fellow who had to sit through the decapitation of Wells to get the collar to the lab. "We have a problem," he said, obviously in an excited state. I was unsure if I really wanted to know.

The FBI was just informed that the owner of the residence next to the Tower Site called the authorities to let them know he had a body in his freezer. I don't recall exact conversations from over a decade ago as a general rule, but I do recall my exact response to this information: "Go fuck yourself, Todd. That isn't funny."

He assured me of his sincerity, repeatedly, and filled me in on the story. Investigators flocked to the scene, where they found both the body in Rothstein's freezer, as promised, and a suicide note. The first part of the suicide note informed the reader that Rothstein had nothing to do with killing the man in his freezer. Had the missive ended there, we would have kept a clean delineation between the two cases, and perhaps taken even longer to link them. However, the last line of the note read, "This has nothing to do with the Wells case." The body in the freezer was that of a James Roden, known from that point forward in our investigative circle as Victim Number Four. The interrelationship of these two cases would be debated for years.

Enough facts emerged in the early stages of the murder investigation to paint a Victorian-era melodrama. How, exactly, did Roden end up in Rothstein's freezer? As it turns out, Roden had been shot by a former girlfriend of Rothstein, Marjorie Diehl-Armstrong. She called Rothstein to help clean up the crime scene and deal with the body. Being a civic-minded man, he obliged. To buy some time, he purchased a freezer to stash the body in while they came up with a plan to dispose of the corpse. It is important to realize that all of this occurred *before* the Collar Bomb incident. This set the stage for what was my initial assessment of events.

Although I could not prove it at the time, I had a personal theory about why Rothstein eventually cracked and called 911. Remember, he lived right next to the site where Wells was abducted. Police were knocking on his door, and, with a corpse in his possession, he was trying to stay below the radar. I can only imagine the stress that Rothstein was under every time he saw police cars going past his house in search of more clues while he was storing a homicide victim in his freezer. In a scenario analogous to the plight of the main character in Edgar Allan Poe's "The Tell-Tale Heart," he was eventually driven mad by the pressure.

To Rothstein's credit, if that is the right way of looking at the situation, he retained his composure for weeks while the Wells investigation carried on. The event that finally broke him down was Armstrong's insistence that they dispose of the body. Armstrong was pressing Rothstein to get an ice crusher to render the corpse into smaller segments that would be easier to deal with. It was this pressure that proved to be the last straw for Rothstein's already frayed nerves.

You might be asking yourself how long it takes a human body to defrost. Most likely, the question never entered your mind before now. However, since I know the answer, I will share it with you. It took three days before investigators could defrost the body enough to move it from the freezer. Every time my wife complains about how the Thanksgiving turkey is still frozen after a day thawing in the fridge, I can't help but do the mental comparison. Hazards of the job, I guess.

Investigators in the Wells case were already dealing with one of the more bizarre bank robbery/murders in US history; no one wanted the corpsicle heaped on top of the quagmire already in their lap. But forensics is a harsh mistress.

Although we could generate no useful leads in the Wells case, the lab was able to notify investigators that lead shot found on the body of Wells was consistent with the shotgun round pellets used to shoot Roden. We had a clue, but it seemed too bizarre early on to believe. No other investigative leads could be found to either separate the two cases or give stronger cause to link them. This stalemate persisted for months.

Early forensic exams had left us with two phone numbers we could not track down and a link to a body in a freezer we could not dismiss. Science did not seem to be interested in working on our side.

With the label of major case came incredible visibility throughout the FBI. It has been my experience that the higher in the ranks your case gets attention, the more potential for both good and bad fallout. A case with the attention of the FBI director is fabulous for all concerned when the big break occurs and the bad guys are captured. Such a case with "executive interest" where every turn hits either another dead end or a multitude of incredible side stories—such as the frozen corpse—garners another type of attention. No one wants this type of scrutiny.

In a typical case, there is a standard number of forensic tools the FBI Lab brings to bear to generate potential clues. After those techniques are exhausted, everyone concerned shrugs his shoulders, packs up the evidence, and moves on to the next case. COLLARBOMB was too bizarre to allow such a graceful exit. The lack of investigative leads put even greater pressure on the lab to come up with any kind of insight that could open up new avenues to probe.

Forensics is a tricky thing. Anyone who watches the news knows that forensic science is under assault. Some of the allegations contain truth; others are grossly exaggerated. The US legal system is adversarial by design. Any forensic technique brought into the courtroom must be robust enough to withstand intense scrutiny. No one wants junk science injected into the legal system. It is for this reason that places like the FBI Lab have a standard suite of accepted techniques that can be solidly applied when examining evidence. Examiners are cautious about straying outside those well-defined strictures. In major cases, scientists will conduct examinations, which, although rooted in sound science, can never be brought into court. Such exams are designed solely to give investigators information that might better focus their efforts.

No one ever knows what clue will be "the clue." If the traditional suite of forensic tests comes up empty, the search for "the clue" can take on more drastic approaches. The history of bombing cases is filled with just such extraordinary examinations.

In July 1996, Eric Rudolph placed what is believed to be the biggest pipe bomb used in the United States to date in the Centennial Olympic Park. The device contained a thick metal plate to optimize the focus of fragmentation. Ironically, this plate ended up saving lives rather than increasing the device's lethality.

Rudolph placed the device in a military backpack under a park bench. Two young men noticed the abandoned backpack and decided it would be an ideal souvenir to take home with them. Upon trying to lift the backpack, they discovered it weighed far more than either had the ambition to carry, thanks to the massive plate and pipes. They dropped the bag back down and walked away. When they dropped the bag, the massive plate tipped the bag over in a manner that directed the pounds of masonry nails wrapped around the bomb skywards instead of straight at the massive crowd attending the celebrations. I can't tell you how many stories of fate intervening I have encountered in my career, but this one has always stuck with me.

The Centennial Park bombing became FBI major case CENTBOM. As this was a bombing against a US Olympic venue, it had much greater ramifications than COLLARBOMB. It, too, called for forensic techniques well out of the norm. The plate from the device turned out to be one of the more unique items recovered post-blast. Insight into its relevance would have to wait many months.

In January 1997, the presence of a metal plate in a device utilized against an abortion clinic in the Sandy Springs Professional Building immediately caused investigators to view the two devices as related. Detailed metallurgical examinations were conducted on this plate far beyond those that would be used in a court of law. The same examinations were conducted on the plate from the Olympic venue bombing. Both plates shared unique physical attributes and an indistinguishable chemical makeup.

The lead FBI metallurgist, who also assisted on the Collar Bomb case, went to all of the steel manufacturers in the United States and collected records of the chemical makeups from all the steel runs they had made over a multi-year time frame. In total, seventy-eight thousand records were collected. In a

Herculean forensic investigation, investigators discovered that both plates possessed a chemistry that was matched by only one of the seventy-eight thousand steel runs. This steel was produced by a specialty steel manufacturer, and a small lot had been sent to one machine shop in North Carolina. It later turned out that Eric Rudolph had acquaintances at this shop, was seen lurking around the business, and had, indeed, stolen the plates from this location. This fact gave investigators the connection necessary to link Rudolph to the Centennial Park device.

Obviously, not every FBI case can take such extraordinary steps in forensic examination. Larger cases justify more extreme measures. Any FBI official "major case" can find itself in the position of needing to throw a couple Hail Mary passes.

In COLLARBOMB, the pressure never seemed to let up. Field investigators got desperate for leads. In an effort to generate new information, the FBI pursued options that they'd not normally consider. They approached the media with select pieces of evidence to share with the general public. They gave *America's Most Wanted,* a television show with a tremendous national outreach, exclusive access to case details and photographs of the collar, gun, and collar box frame to televise to see if anyone recognized the items. Over the course of the case, *America's Most Wanted* aired five episodes featuring COLLARBOMB.

These high visibility TV shows not only kept COLLARBOMB in the forefront of the public's mind and view, but also kept the case front and center in the minds of all the FBI leadership. Both the field and the lab were kept under constant pressure to come up with the lead to break the case open.

Throughout the case, we uncovered a few unique items that led to many searches. For instance, with multiple tool marks on the components used to fabricate the device, it was obvious that common tools such as screwdrivers and drills were used to build the bomb. They left behind distinct, valuable markings. So, in every search conducted during the investigation, tools like screwdrivers and drill bits were collected.

At one point in the case, it became apparent that Rothstein—the man who left the suicide note proclaiming no association with Wells—was most probably the bomb builder. Investigators discovered that Rothstein had a close acquaintance with an extensive machine shop. A search warrant was issued for his friend's shop. I joined the investigators and evidence recovery team to assist in the search.

There are people who are packrats and there are people who are thieves. These two demographics are not mutually exclusive. Rothstein's friend had worked at various locations over the course of his professional life and had access to a wide variety of tools. His machine shop was packed with a lifetime's worth of tools, which appeared to have been liberated from these locations. During the course of this one search, our team collected 120 screwdrivers and 640 drill bits of a wide range of sizes. We also found various metal plates similar to those used in the assembly of the collar and metal box that housed the bomb.

Our tool-mark examiner spent hundreds of hours—approximately six months—painstakingly comparing all the tools we collected over the course of the case and never found a single match. It's a wonder he did not shiv my technician with one of the tools.

Traditional forensics also yielded unique hairs and fibers of interest. Within the duct tape collected from the device were dog hairs and green carpet fibers. The fibers were unique enough that the examiner was convinced a strong association could be made if we ever found the source. These two pieces of information led to the FBI collecting dog hairs from anyone even remotely associated with suspects. The canine source was never located. In very few instances did we ever encounter green carpets, but, in the one or two cases they were encountered, fibers were collected. None ever matched.

As traditional forensic clues were exhausted, less common analysis was pursued. The collar itself had a plastic tube filled with blue fluid laced through it. Normally, an analysis of such a fluid would only be undertaken if something to compare it against was found. For example, if during a search a vial of blue fluid was found in a suspect's workshop, it would be compared chemically against the blue fluid in the collar to see if the two could be linked. Such a link helps to build associations between the suspect and the bomb. However, we never found blue fluids in our searches.

As other leads started to die off, we eventually conducted an analysis on the blue fluid. Its constituents led us to believe it was some sort of hydraulic fluid. Investigators, in need of any leads, started to go to the local automotive parts stores in Erie and collect samples of hydraulic fluid from the shelves. A match with one of these products might open a lead. Again, no one ever knows when and if "the clue" is going to appear. Sometimes you have to try and force it to show. None of the fluids ever matched. The only thing that it produced was another frustrating dead end.

One of the amazing things about working high-profile bombing cases is I get to see many of the FBI's best assets deployed. When field investigations fail to turn up leads, and forensics proves equally as fruitless, the FBI has yet another ace up its sleeve: the Behavioral Analysis Unit (BAU).[23] Most folks refer to these specialists as the "profilers."

[23] Forensic scientists are not the only ones who have their profession bastardized by TV and movies. With the introduction of *Criminal Minds*, our brethren in the BAU now share our pain. I have watched just enough of the show to see the heroes riding all over the country on the FBI Gulfstream to save the day. Having been on missions using a real FBI Gulfstream, I can personally attest that this aircraft is not open to such joy rides.

I am a chemist, an explosive engineer, and sometimes a demolition specialist (if no one more capable is on the range). In no capacity do I profess to be a behavioral specialist, profiler, or psychologist. However, as the Collar Bomb case continued to stymie everyone struggling to gain answers, the FBI inevitably turned to the BAU team. They, in turn, came to me and asked for my assistance in preparing a profile of the bomber(s).

It is a running joke amongst some FBI agents that if you ask BAU for a profile of a perpetrator, they will come back with a statement that the suspect will be a white male between thirty-five and fifty-five with anger issues and a dislike of authority figures. Since I oftentimes skirt that demographic, I treat it with some skepticism.[24] However, in a case where any insight can potentially add to a dearth of leads, and perhaps open up new avenues of pursuit, engaging our profilers was worth the time.

Psychologically, we actually had a treasure trove of information to pull from. The bomber(s) had provided eight pages of detailed written instructions to the bank and to Wells. These instructions were packed with vitriol, editorial diatribes, and what could only be described as rants peppered with delusions of power. Some of the pages can be found online, with text consisting of:

> *"If you delay, disobey or alert anyone you will die! It is your choice to live or bring death."*
>
> *"Stay calm and do as instructed to survive. We're following your moves in cars to make sure you obey. Sentries are driving and looking out for authorities. We are scanning police radio frequencies and cell phone calls. If police or aircraft are involved, you will be destroyed."*

---

[24] Don't get me wrong, I have a great deal of respect for the skill set of my BAU colleagues. I just wish they wouldn't pick on us angry old white guys so much.

The BAU profilers had ample red meat to develop a personality type from the notes. What they wanted from me was an evaluation of the type of person who would build the bomb. Over the years, I have had the chance to work with many law enforcement agencies across the world. In this capacity, I have heard the motivations behind many acts of barbarism. I have seen a wide variety of explosive devices. Most have a basic utilitarian aspect to them. In other words, most bombs are put together with the express purpose of blowing up with no fancy trappings beyond what is required to accomplish this very direct goal. The collar-bomb device was different.

This device oozed hatred. I am not allowed to make such statements on the stand, as I cannot infer the intent or motivation of a bomb builder. An agnostic analysis of wires, switches, and pipes does not allow for inference of the bomb builder's psyche. COLLARBOMB was the only case in which I felt I could almost gaze into the internal workings of the bomb's creator. Working with the profilers, I shared my interpretations of the person who assembled the bomb.

The bomb was built with the same malice as flowed off of those pages. Every part was over-engineered and designed to meticulously fit a designated purpose. It was my feeling it was meant to kill the wearer. The best analogy I can give is to the movie franchise *Saw*. Devices produced by the killer (Jigsaw) in the movie are obviously created by Hollywood special effects artists. These artists create elaborate devices designed to look menacing, powerful, and evil. If Jigsaw was to design a bomb, it would have been the collar bomb.

In many ways, the collar bomb was a work of an artisan. With the exception of commercial constituents like kitchen timers, many pieces were custom made for very distinct purposes. No one who went to the trouble to put weeks into designing and fabricating such a bomb would have no previous history of fabrication. In addition, this same person fabricated a home-made shotgun using many of the same materials as put into the bomb. This was the type of person who had been making things for years. Over-engineered things. Weapons. The type of crap he would have to show off to others. The type of thing that, when others saw, they would immediately start wondering what sort of person builds those types of things.

And, so, the profile developed. Many aspects of the personality laid out in the profile matched the group of miscreants eventually rounded up over the seven-year investigation. However, like with forensics, the BAU profilers' excellent work did not break the case open.

Putting back together a bomb is rarely simple. Imagine purchasing a one-thousand-piece jigsaw puzzle. Next, take half of the pieces and throw them away. Throw gasoline on the remaining half and light them on fire. Finally, beat out the resulting inferno with a tire iron. My job is to collect these pieces and figure out what the original picture was supposed to be.

I can tell you that the bomb was housed in a box made of angle iron. Pictures of this box abound on the web, as they were released as part of the first *America's Most Wanted* episode, as were pictures of the collar. The front of the box had an opening covered by glass and a wire mesh. There were numerous warning labels on the outside of the box. The bomb builder not only sprinkled warnings about the bomb throughout the notes given to Wells but also posted these threats all over the bomb. A litany of cautions, ranging from hidden tripwires to a whole host of actions that could set the bomb off, covered the exterior of the casing.

Contained in this box were the device and the fuzing system. The actual bomb consisted of two steel pipes filled with smokeless powder, the exact same powder found in shotgun shells. The most common explosive fillers used in bombs in the United States are black powder and smokeless powder. This is because they are easily purchased and bombers tend to be lazy. Metal plates were used to seal the ends of the pipes. When smokeless powder burns under confinement, it produces extremely hot gases under high pressure and fractures the pipes. To this day, I still do not know the exact method used to ignite the explosive in the pipes. In most bombs, a thin filament, such as that found in a light bulb, is inserted into the explosive to create the heat needed to set it off. Such filaments can be very small, fragile, and very hard to find after a bomb goes off.

The fuzing system in the bomb used two kitchen timers, two AA batteries, and numerous switches. I estimated the device contained three separate switches. One of the switches protruding from the back of the metal box was thrown to arm the bomb. The next switch consisted of two modified kitchen timers. The timers were fitted with metal rods that made contact as the timer face swept the clock and came back to zero.

In addition to the obvious fuzing components, two additional items were recovered. These were the toy cell phone and electronic countdown timer mentioned earlier. Over the first month, we were able to determine the cell phone was not incorporated into the workings of the bomb. It was designed to be visible from the outside to give the appearance of a remote form of initiation. The digital timer was also found not to have been altered. The closest use I could come up with for the timer was to give an outward indicator of how much time was left before the bomb blew up. The bomber put the device around Wells's neck and then switched on the timers to start ticking down to zero. Wells was supposed to deliver the money to the person(s) who strapped the bomb to him. I imagined the bomb builder would have liked to have some way of knowing for sure how much time remained before it would be unwise to approach Wells. Obviously a person could start a personal timer, but this builder wanted the added complexity in the bomb itself. It also struck me that if more than one conspirator existed, the timer allowed anyone who was part of the plot to know the time remaining. Remember that the last thing witnesses heard before the bomb exploded was the beeping of the timer. True to form, it kept track of the remaining minutes in Wells's life.

Everything I have described so far was fairly obvious within the first weeks of putting the bomb back together. Also present in the debris were hundreds of assorted fragments that basically made no sense at all at first glance. Small sections of Masonite and fiberglass reinforced panels (FRP), like those found in many shower walls, were scattered throughout the evidence. Also present were metal rods, bolts, springs, wire mesh, and small locks. Each of these served a potential purpose—one that was in no way obvious.

I, along with Rex who was a more seasoned examiner and whose mechanical abilities surpassed mine, took to rebuilding the bomb. It literally became the jigsaw puzzle, but in 3D. The puzzle aspects of this process were fascinating and frustrating all at the same time. I had 120 bags with a random assortment of crap distributed throughout them. Just looking at the Masonite and FRP sections alone, dozens of pieces had to be fit together. I had to keep track of which bag every piece came from. When possible, we wrote numbers on pieces. Many were too small for such labeling.

For these pieces, we took detailed photographs and notes to account for their bag of origin.

It took Rex and me two stressful months of puzzling out pieces to finally get most of the device put back together. At times we came fairly close to blows. I remember telling him I would be better off breaking out the Ouija board for guidance than taking one of his conjectures. Following his retort regarding the integrity of my lineage we took an hour off to refresh our perspective. Even after months of work we could not determine some of the device's functions. After some lengthy discussions, we decided a fresh eye was required. We presented the field bomb tech who was involved from the very beginning with a "professional growth opportunity." As this was the same guy who had had to witness a decapitation, I am still fairly surprised he took us up on our offer.

We took the bomb apart and had him come to the lab for a month. We told him it was his turn at the puzzle. Without telling him anything about our conclusions, we put him in the room with the evidence and told him to rebuild the bomb. Two weeks later, he had basically assembled the components in the same fashion but figured out the location of one additional piece (a small luggage lock) that had eluded us since the beginning.

One of the attributes that made the collar-bomb device so unique was the custom-built nature of many of its inner workings. It turned out that the Masonite and FRP sections were originally put together as a custom holder for the two kitchen timers. Underneath the base of this holder, a groove was cut. This groove fit a section of wire that was attached to two bolts. A thin metal rod with metal attachments on each end sat on top of the timers in a recess carved into the Masonite. Metal bolts and brackets held other pieces securely in place. We now had the pieces mostly situated, but in some ways that was the easy part of the puzzle.

With the device finally assembled, I had to figure out what all the pieces I had struggled to put back together actually did. Some, like the timers, were obvious. Others were baffling. Had I had the whole device in front of me in pristine working order, I could have set things in motion and watched how all the bits interplayed with one another. Instead, I had a cobbled together mound of fragments that were equal parts scotch tape to evidence.

On a case this complicated, everyone in the explosive unit becomes engaged and puts forward ideas. Slowly, ever so slowly, a coherent picture started to emerge. The one piece of evidence put back in the puzzle by our field bomb tech shed light on the purpose of the mysterious rod. The automotive expertise of Rex provided another startling revelation. Mark, his house painted and sold, was also a source of valuable insights. About five months into the examination, we had our bomb back together and a fairly good idea how it functioned.

Without providing a bomb-making tutorial, here is how the bomb functioned. After Wells had the bomb strapped around his neck, one of the suspects pulled a key ring attached to a cotter pin on the back of the device. Imagine pulling the pin on a hand grenade. The construction was very similar. Except, in the case of the collar bomb, the pin held in place a steel bolt under spring tension. The moment the pin was pulled, the bolt slammed into the inside of the device and made contact with a metal plate. This action armed the bomb. In theory, someone could have de-armed the bomb by pulling this bolt back out, but it was flush with the back of the box.

Careful inspection showed that the bolt's shaft had been drilled and threaded. This would have allowed someone to screw in a smaller bolt into the bolt shaft to pull it back out. However, a bolt of the proper size and thread profile would be needed. At this point, the bomb was armed and ready to be put in motion.

On the front face of the bomb was a panel with multiple cotter pins attached to key rings. To start the countdown, the bomb builder pulled one of the pins out. This action freed one of the kitchen timers to start its countdown. In theory, someone could have put a safety pin in the hole left by removing the cotter pin.

The bomb builder thought of this. On the inside of the front panel were small metal plates under spring pressure. The minute the pin was pulled to release the timer, the spring pushed a metal plate to cover the hole from the inside, blocking all access. The bomb was now live and counting down. Wells had approximately fifty minutes before the bomb would explode.

The bomb maker anticipated there might be a need to either kill the bomb or delay its explosion, depending on how the plot unfolded. Mechanisms for both potential courses of action were included in the device. The secret to killing the bomb lay with the miniature lock whose placement our field bomb tech figured out. The holder of the key to that lock could have inserted the key into the side of the device and opened the lock. By opening the lock, the mysterious metal rod would have been freed. At the end of the metal rod was a piece of metal covered with epoxy. The freed rod would slide out of its recess and push the epoxied block between the timer dial and the screw it was set to make contact with. This action would have stopped any chance of the bomb going off.

The numbers "57/50" were written on the front of the bomb box in red marker. These numbers taunted me for a long, long time. As the device came together, the shroud was lifted on their meaning. The first timer was set for fifty minutes. When the first cotter pin was pulled from the front face panel, a fifty-minute countdown began. If you recall, there were two kitchen timers in the bomb. There was also one cotter pin still inserted in the front panel of the device. When the entire bomb was reassembled, it became obvious that this second cotter pin controlled the second timer. All the while Brian Wells was sitting in the parking lot waiting for the bomb squad to arrive, the means of delaying his demise was literally right in front of him. The bomb builder constructed the device so that pulling the second cotter pin would free the second timer. This would create an additional fifty-seven minutes' delay. If Wells knew this, he never told the arresting officers. But it is doubtful he was aware a mechanism to extend his time was in his grasp.

In addition to the two safety mechanisms, the box held one cruel trick hidden in its depths. To this day, I am unsure if it would have worked, but all indications pointed to a hidden booby trap. As described earlier, the front of the box had a window that allowed officers to see a portion of the bomb's inner workings. The window was covered with glass, and behind the glass a wire mesh provided a secondary covering. One red wire was hooked to this mesh. To discern what this wire did would require all the remaining wires to be intact. Bombs do not leave things intact when they go off. I was able to find enough sections of wire to determine that the mesh was indeed hooked up to the bomb's power source. It is my belief that anyone trying to break into the bomb through the glass would have set off the bomb the second that the wire mesh touched any part of the box containing the bomb. Such a booby trap is rare and stood as further testament to the ingenuity of the builder.

There is one attribute of the device that spoke of intention above all others. This was the solid metal backing plate used in the box that housed the bomb. This is the plate that was resting closest to the chest of Brian Wells.

As a forensic examiner, I cannot speak with 100 percent confidence to a bomber's intent. Steel, wire, and explosives don't provide insight into the inner workings of a bomber's mind. However, some aspects of a bomb's construction can be used to infer intent. For example, if someone adds a pile of steel ball bearings to a bomb, I can infer he intended to create additional fragmentation. As a scientist, I know that fragmentation increases the lethality of a bomb. While I can't specifically state that the bomber added the ball bearings with the intent to make the device more lethal, I can state that the effect will be such. The jury needs to make the final evaluation for the bomber's actions.

The metal plate resting against Wells's chest was intentionally scored in a checkerboard-like pattern. There is no earthly reason to alter a sheet of metal like this except for one purpose. The bomb builder wanted the metal to fragment into pieces with the explosion. Early grenades had the same type of scoring on their interiors to create reproducible fragments. In the case of the collar bomb, this plate was not facing outwards towards the world; it was facing directly at Brian Wells's heart and lungs. It was the force of this plate driven into his chest that produced the fatal injury. While I can't state so on the witness stand, my opinion is that the bomber designed this device to optimize the chances it would kill its wearer.

In many ways, the Collar Bomb case is not a victory story. Those interested in the investigation are invited to read my friend Jerry's book, *Pizza Bomber*. Jerry Clark's outstanding investigative work and years of patience eventually uncovered the conspirators behind the elaborate bank-robbery plot. Over the years it took Jerry to build his case, the crew who pulled off this crime slowly started to either die off or plead to other charges. Most frustrating to me was the fact that Rothstein eventually died of cancer. He was the main suspect as the builder of the device—something to which he never admitted, even on his deathbed. I will never know who constructed the most convoluted bomb I ever examined.

At the end of the case, only one suspect was left to take to trial: Marjorie Diehl-Armstrong. She was, to be kind, petty evil incarnate. She was also eventually found guilty of conspiracy to commit armed bank robbery; armed bank robbery in which death resulted; and use of a destructive device in furtherance of a crime of violence. As some closure, I was able to testify in her trial, even if my testimony was fairly limited to how the collar bomb was constructed and how it functioned.

The collar bomb also had one other unique aspect, which, in the end, frustrated my ability to even fully do my job in court. In most cases, it is my job to explain to the jury the potential harm such a device can do. Oftentimes, I have to reproduce a device on a range to film its destructive power. However, the collar bomb explosion was actually caught on tape.

As Brian Wells sat on the parking-lot asphalt handcuffed and surrounded by police officers, a local news crew arrived. They were kept at a distance by police but still had line of sight of Wells in the lot. Cameras were rolling as events unfolded. In one video clip, a reporter talks to his cameraman with Wells clearly visible a short distance away over the reporter's shoulder. The reporter is explaining how local police have told them to keep away. He notes that "we should be safe here." At that moment, the bomb strapped to Wells explodes in clear view of the camera.

There are some historical examples of reporters capturing moments of tragedy in our culture and clearly being affected by the events unfolding around them. As the *Hindenburg* went down, the reporter's cry of "Oh, the humanity" resonates with us all. In the case of Brian Wells, as the reporter spins around after the explosion and sees Wells thrown backwards against the hard ground, the only words out of his mouth are, "Did we get that?"

Some days, my optimism for our species wavers.

With a video of Wells being killed by the bomb, there was very little I needed to tell the jury. They were shown videos of the event, and pictures of the hole punched into his chest by the metal plate I spoke of earlier. My contribution was to let them know that this device was designed in a manner to optimize its capability of killing the wearer and give them a feeling for how its function fit into the larger bank-robber plot.

COLLARBOMB took up a good portion of my professional time as a forensic examiner. In the end, forty-six separate batches of evidence from searches were sent to the lab as part of this case. Within these submissions were 214 items from the crime scene and 1,289 items collected from search scenes. Every forensic unit in the FBI Laboratory touched the Collar Bomb case. Fifty-eight separate forensic reports from eight major forensic disciplines were issued over the seven-year span. Thousands of hours of time were spent by very skilled and extremely over-tasked forensic examiners.

I am conflicted on how I feel about COLLARBOMB. A good part of my job revolves around attempting to thwart terrorist bombings. In no way, shape, or form do I want anyone to think I have an admiration for those who would do harm with bombs. However, there is more of a sense of purpose to the time spent fighting an adversary such as a terrorist. It feels like playing a role in a larger, timeless struggle of good against evil.

The perpetrators of the Collar Bomb plot were petty, egotistic nothings. There was no grander scheme behind their actions, either perceived or ethically debatable. Spending years working on a case with such base motivations adds insult to injury in a very odd way. There are truly dedicated people intent on using violence to take down the very way of life of good and free people. Focusing energy working on the actions of greedy lowlives detracts time from fighting a greater evil that desperately needs to be addressed. In the end, that leaves me more frustrated and conflicted than rewarded, regardless of the outcome.

CHAPTER 10

A House Divided

When I began my career investigating bombings, I had only a surface level of understanding of the underlying political unrest that served as a catalyst for bombing attacks. But as I visited countries that had decades of unrest and political distrust, I started to understand more—and how this history traps generations.

Throughout the '90s, while at New Mexico Tech, I worked with scientists and bombing investigators from the United Kingdom, focusing on countering the massive vehicle bombs the Provisional Irish Republican Army (PIRA) used to target the UK mainland, mostly London. During the many tedious hours of making tons of explosives to test, I was regaled with countless tales of the atrocities PIRA bombers had committed. Being a young scientist and researcher who had only recently ventured away from the East Coast of the US, I could not envision what would drive someone to commit such horrendous acts of violence. But a decade later, a chance trip opened my eyes.

In 2009, I attended an international meeting of Bomb Data Centers (the groups of people who keep bombing stats for intelligence and law enforcement purposes for their nations). The gathering was held in Belfast, Northern Ireland. One afternoon, some colleagues and I decided we wanted to see—really see—Belfast.

It is the mark of the human spirit that people find a way to live with tragedy, and sometimes make a living in its wake. In Belfast, you can take what is referred to as a "Bombs and Bullets" tour. Our group asked our guide to see the "murals" and highlights of the conflict. To this day, I still can't wrap my mind around what we were shown.

Murals depicting tributes to "victims" and "heroes" of the conflict on both sides covered whole buildings. Some were moving tributes, others unnerving promises of retribution. All represented the hopes and aspirations of two sides locked in a conflict that neither seemed able to leave behind. I saw whole neighborhoods locked down behind twelve-foot concrete walls topped with barbed wire fences. The walls served as barriers, artifacts of an engrained hatred, to either keep an adversary out or lock in and control the same.

Over the course of the conflict, PIRA had made a habit of targeting the police with mortars. To protect police stations, authorities built netting around the lower floors to deflect projectiles. Unfazed, PIRA made their mortars stronger, to be able to lob their bombs over the fencing. On our tour, we passed a four-story-tall police station. Chain-link fence surrounded the entire building, including the roof. The building was literally encased in fencing.

What I could not comprehend as the newcomer to the field in New Mexico started to make tragic sense in Northern Ireland. I am not saying I understand, or in any way condone, the violence. Visiting these sites, however, I could see how people stuck in such circumstances could get to the point of hatred. It was a sobering moment.

At one point, we ended up in the PIRA graveyard. No words can describe the weight of the place. Six-foot-high pedestals, adorned with an assortment of equally massive angels and crosses, served as grave markers. These monuments went on as far as the eye could see on the rolling hills. The cab driver took us to the heart of the graveyard where the hunger strikers from 1981 were memorialized. Amongst the hunger strikers was their young leader, twenty-seven-year-old Bobby Sands, who would die after sixty-six days of self-starvation. We stood at the foot of his grave in silent contemplation. At this point, our driver noted that a funeral procession was arriving fairly close and it would be prudent to "leave the area quickly." The insights I gained during that tour of Belfast put into perspective the decades of stories I had amassed on "The Troubles" in the area.

As a bombing investigator, there are two places that need to be visited. One is Northern Ireland. The second is Israel. I had visited Israel years earlier on a high-profile case. With Belfast behind me, Israel, in hindsight, also made a great deal more sense.

On October 15, 2003, at approximately 10:15 AM, a massive bomb buried near Beit Hanoun Junction in the Gaza Strip detonated while a convoy of vehicles from the American Embassy passed by. The three-car convoy of armored SUVs escorted by Palestinian police was heading south on Gaza's main road just after entering the Gaza Strip from Israel. The first two cars passed by the bomb, and, just as the third car came into proximity, it was detonated. Electrical wires were found at the bomb site and led to a nearby building, which served as a strong indication that the bomb was manually detonated, with the US convoy as the intended target.

Embassy vehicles are fitted with a degree of armor designed to ward off attacks from small-arms fire. While the steel reinforcements are sufficient to stop small masses of lead traveling at upwards of 1,200 feet per second, withstanding the blast from forty pounds of high explosive is another matter. The bomb that blew up next to the American convoy was a "tank killer"; the smaller Suburban didn't stand a chance. The resulting carnage was horrific by any standard. Mr. Mutya Abdel-Wahad (a nearby shop owner) stated through a translator, "I was the first arrived, and I saw half a body laid on his face and another half a body like thirty meters from the explosion."

In its wake, the bomb left a crater three to four feet deep in the unpaved stretch of road. The car passing the device was flipped over and torn in half. Two of the occupants were killed immediately. A third died on the way to the hospital. The fourth survived. Americans, foreign journalists, and Palestinian security men who tried to respond to the bomb site that day were pelted with stones by Palestinian youths. The violent attack forced the Americans to retreat into their cars and rapidly depart the scene.

All four victims of the attack had been hired to provide security for US Embassy staff. The men who gave their lives while performing these duties were John Branchizio, thirty-seven; Mark T. Parsons, thirty-one; and John Martin Linde Jr., thirty.[25] I wish I could have done more to bring their attackers to justice, but this chapter is dedicated to their memory.

The attack against the American convoy occurred slightly more than three years into the second *intifada* (series of Palestinian protests). It marked the most lethal attack ever directly

[25] The name of the fourth victim, who survived that horrific attack, was not released by the State Department. I hope he successfully moved on from the trauma of that day.

targeting US personnel in Israel, the West Bank, or Gaza. While, historically, Palestinians often accused the United States of siding with Israel, a line was drawn against blatant violence targeting our country and its representatives. Palestine Liberation Organization (PLO) chairman Yasser Arafat and Palestinian prime minister Ahmed Qurei quickly condemned the bombing. The main Palestinian terrorist groups (Hamas, Islamic Jihad, and al-Aqsa Martyrs' Brigades), who were responsible for countless deadly bombings against Israelis, also publicly distanced themselves from the attack.

Sadly, the bombings diminished the real, good purpose for the convoy of embassy personnel that day. Diplomats in the convoy were heading to Gaza to interview Palestinian academics seeking Fulbright scholarships, which would allow chosen candidates to teach or study in the United States.

With the death of three American citizens, there was no doubt the US government would want to lead the investigation into the Gaza bombing. In any regular attack overseas, this would be tricky. The US does not own a crime scene that happens outside its soil. The only exception to this is an attack against a US embassy, specifically, the grounds the embassy stands on.

Had this been an attack in any other country, simple diplomatic negotiations would clear the way for an FBI team to come in and collect evidence. This, however, was not just any other country. This was Israel, one of the most disputed areas in the world. To make matters worse, this was an attack that occurred in Palestinian territories within Israel. There was no real precedent for how to move forward.

The first US team who went to examine the scene were greeted by angry mobs and flying stones. As a result, the Palestinian

Authority collected much of the evidence needed by the FBI. This meant going back into the Gaza Strip to retrieve it.

Historically, Explosive Unit forensic bombing examiners were selected from a cadre of FBI special agents. These agents, typically FBI bomb technicians, were brought back to the FBI Lab and taught the rigors of forensic examinations and testimony. That changed in the late '90s, and a New Mexico Tech colleague (Michael Leone) and I became the first scientists to join the FBI in this capacity. I would later become the first non-agent[26] to become a fully qualified Explosives and Hazardous Devices Examiner for the FBI.

For the most part the duties of agent examiners and non-agents such as myself were identical. Only in very specific circumstances would any differences emerge. The initial entry into Gaza was just such a case. The FBI decided that anyone going in with the Explosive Unit team to collect evidence would have to be armed for his own defense. As I was never issued a gun—and public safety is best served with me remaining unarmed—I could not participate in the first incursion into Gaza. A close colleague of mine, Rich Stryker from the Bangladesh chamber of horrors, did travel to Gaza, and he witnessed Israeli fighter jets flying overhead to conduct air strikes in retaliation for the bombing.

At the time, I was at a work conference in South Africa. Upon my return to the States, I learned that the forensic examination of the Gaza case had been assigned to me. I remain unsure how my number came up for this case, as I was still dealing with the fallout from the Collar Bomb investigation. Needless to say, it was shaping up to be a very busy year.

---

[26] I beat Leone to this honor by mere months because I had to accelerate my training to allow me to testify against Ahmed Ressam (the Millennium Bomber) with respect to testing I had conducted to assist the FBI while still a researcher in New Mexico. No good deed goes unpunished.

When I opened the evidence for the Gaza case, one thing was clear to me: I had very little to examine. When a bomb goes off, a great deal more survives than would be expected. For example, in the Collar Bomb case, I had hundreds of small pieces that required reassembly, but it was a possible task. In Gaza, the bomb was buried and held upwards of forty pounds of high explosive charge. A device that size requires a substantial container. In theory, there should have been lots of shattered pieces to put back together. However, there are only lots of pieces to examine if, and only if, the folks on scene actually collect them.

In the immediate aftermath of the bombing, the targeted Americans' only concern was getting out of the territories and getting the survivors medical attention. Collecting bomb fragments was not top priority. The team that tried to come back to the scene collected more hurled stones than evidence. So the collection was basically left up to the Palestinian authorities. For all I know, they meticulously picked up every single piece of the device and cataloged them with great care. If they did, the full collection of evidence and its documented inventory wasn't what they handed over.

I had enough components to figure out that the bomb was command detonated. A very long length of wire attached to a battery connector showed the device was hardwired. The buried wire was traced to a nearby building with line of sight to the target. Left behind was a cache of nine-volt batteries hooked together, which obviously powered the device. Finally, a switch was wired into the circuit to allow the bomber to select the exact moment the bomb went off.

A known fuzing system fills in a lot of blanks for investigators. At least we had that. From an investigative standpoint, the items we had indicated that the bomber meant to hit the American convoy. Those SUVs stand out against a sea of miniature cars like elephants in a kiddie pool.

However, I still did not have enough pieces of evidence to figure out anything beyond the fact that the bomb was housed in some sort of metal container. Early in November, a few weeks after the bombing, we caught a break. During a raid by Palestinian authorities at the residence of a potential suspect, a full bomb was recovered. Investigations led them to believe this device and the few fragments from the one I possessed were related. They wanted us to do a comparison.

To do a comparison, I had to have the bomb back at the FBI Lab in the States. I was fairly certain that a commercial airline would not take kindly to any attempts to carry an IED on board, even with assurances that, most likely, all explosives had been removed. This meant we needed to deploy an FBI aircraft. Luckily, an FBI jet was free. Rich, who had made the first evidence run, and I loaded our gear and headed out to pick up a bomb whose exact makeup we would know nothing about until we saw it firsthand.

Missions like this are a delicate political dance. It is hard enough to get permission to go into some countries. Getting official visas lined up for passports can be a complicated process. In addition, passport visa applications rarely list a reason for visit as "bomb pick up." So, we traveled into Israel and Gaza on our tourist passports. We were instructed to tell no one where we were going, or what we were going to do. No one, not even the president of the United States, who received official daily updates on our status, knew the full story about our bomb-retrieval journey.

At the time of the bombings, Yasser Arafat had control of the Palestinian Authority. He was in a delicate position. The death of three Americans was not going to be ignored by the US. With the FBI conducting all the forensic analysis, Arafat had to show that serious efforts were being made to conduct a criminal investigation in Gaza. To head this task, he appointed a committee of high-ranking Palestinians, among them a colonel from the Palestinian Civil Police and a colonel in the Palestinian National Security Force. My team was slated to meet with this committee; they had custody of our bomb.

Our team arrived in Tel Aviv and checked into the hotel. In many countries, hotels are routinely monitored—and hotel rooms frequently tossed—by foreign intelligence officers. The group noted that three of us checking in were all in rooms directly in line with each other: rooms 314, 414, and 514. I initially attributed it to an odd coincidence; later, I surmised it was the section of the hotel easiest to run video and microphone lines through for monitoring particular guests of honor.

The next day, our team assembled at the US Embassy to strategize our plan to pick up the bomb. Tensions were high between the US and the Palestinians, so the Embassy did not want to send our team too deep into hostile territory. Luckily, a Palestinian colonel had what was referred to as a "guest house" just across the Israeli border with the Gaza Strip. It had walls surrounding it and would be easy to make quick egress in the event that things turned ugly. The meeting with the Palestinian Committee was set up for that location. Now we had some dress code logistics to sort through.

The US had received some intel that Palestinians who were not sympathetic to our cause had been tipped off about our foray into their territory. Rumors developed about potential snipers targeting our crew. We discussed donning body armor to provide a modicum of deterrence to randomly flying bits of lead, a notion then dismissed out of fear of insulting our Palestinian hosts, who had promised our security. Instead, our security detail, who would not enter the meeting room, would wear Kevlar, while the diplomatic team would not. This would not be the last time I would be surrounded by armed security sporting only a light polo shirt and barely contained trepidation.

With the dress code settled upon, our team loaded into two armored SUVs and headed out. The fact that it was these same model SUVs that had been targeted did not escape my mind. I hoped bombs, like lightning, sought out different targets each time.

Located on the north end of the Gaza Strip, the Erez Crossing, or Beit Hanoun Crossing, is the only location for the flow of people between Israel and Gaza. It serves as a choke and control point. During the second *intifada*, the flow of people across this barrier was tightly monitored. Even with our diplomatic convoy, it took almost an hour to receive permission to cross over. An hour spent in boredom can seem an eternity. One spent waiting to drive into potential sniper fire to pick up an unknown bomb still feels like an eternity but is not what I would call boring.

With our drive across the border behind us, I was relieved when we arrived at the "guest house." Our convoy pulled into the walled-off garden area and the armed contingency jumped out of the cars to sweep the perimeter. They did a quick scan of the surroundings, quickly entered and examined the house, and just as quickly came back and told us to hastily get inside. I was not getting any more relaxed.

We were ushered up a flight of stairs to the second floor of the building and into a large conference room. It seemed like we had arrived unfashionably early to the party and had beaten our hosts. Our security detail dropped us off in the room and took defensive positions around the house and grounds. Sometimes, cases don't feel real until something crosses my path and takes hold. In some cases, it's a smell, like decaying flesh and diesel. In Gaza, it was the twenty-foot-long heavy wooden table that dominated the conference room with a six-foot-tall headshot of Yasser Arafat hanging on the wall at its end. Our team stared at the picture for what seemed like an eternity before silently locking eyes with each other. This was our shared "oh shit, here we go" moment.

My eyes went from my colleagues to my surroundings. The venetian blinds over the windows had been pulled down—comforting, I suppose, because it would require anyone trying to shoot into the room to guess where we were. But the blinds on the windows that overlooked the front courtyard looked fairly badly damaged, as if someone had punched holes in them and pulled the plastic slats ninety degrees from the window. The wall opposite the windows was pockmarked with divots taken out of the plaster. I thought immediately of someone randomly hammering the wall to find hidden loot. For the life of me, I could not figure out the pockmarked pattern.

Before I could contemplate the décor further, our hosts arrived. We were greeted by the two colonels and their entourage of helpers. In true diplomatic fashion, their team took up positions on one side of the massive table while we positioned ourselves on the other. The two-dimensional Yasser smiled approvingly at the proceedings. In truth, I don't recall much of the conversation. It was filled with a great degree of perfunctory diplomatic niceties that, over the years, I've learned to tune out. It was with relief that I greeted the security folks who knocked on the door to tell us our bomb had just arrived. Rich and I politely stepped out of the room and allowed the diplomacy to carry on without us.

Even though I was exposing myself to potential sniper fire, it was a relief to leave the conference room and step out into the open.

I had no place in the discussion in that room. Now I was heading out to pick up a bomb. As odd as it may sound, I was entering my comfort zone.

The courtyard in front of the building was fairly open, accessed through a large gate where two cars could pass each other side-by-side. A parking area big enough for a dozen or so cars spanned the front of the yard, and a garden with some thin rose bushes and other ornamental plants flanked the far end of the parking lot.

We were greeted by the Palestinian bomb techs who came with their special delivery. There is an unspoken camaraderie between those who work in my field regardless of nationality. We shook hands and exchanged pleasantries with our colleagues. Through our interpreter, we discussed what they had recovered and various aspects of their experience with the bombs utilized in the area. It provided much-needed insight into what I should look out for in my own exams.

Eventually, the lead Palestinian bomb tech walked around the far side of the building to bring the device over. I expected him to drive around with some sort of SUV, or perhaps a bomb trailer designed to transport explosive devices. I was not prepared for him to come around the corner behind the wheel of what appeared to be a Ford Pinto.

In 1970, Ford introduced the Pinto into the American market, where it flourished from 1970 to 1980. The Pinto had many charming attributes. It was a fuel-efficient sub-compact vehicle sold at a very approachable price. However, there was one unfortunate design flaw with the Pinto. Ford chose to place the car's gas tank at the back, directly between the rear bumper and rear axle. The fuel tank walls were very thin, likely a money-saving tactic in the design, and four poorly arranged bolts held the fuel tank in place. Rear-end collision tests demonstrated that the bolts would puncture the thin walls of the fuel tank in collisions over 25 mph, which often resulted in a gasoline fire.

Somewhere in the range of 27 to 180 deaths were reported as a result of rear-impact-related fuel tank fires. Consumer advocates eventually declared the Pinto "unsafe at any speed." To this day it holds the number one spot on the list of most dangerous cars in history. And at that moment in the courtyard, it was holding my bomb in the trunk, right over its fuel tank.

The Palestinian tech parked the car at the far end of the lot and popped open the trunk. He reached in barehanded and pulled out a massive metal bomb casing. "Now I know at least one set of fingerprints on the bomb will belong to him," I thought. And then: "That is one goddamn big bomb."

The device itself was about the size of an old metal milk jug. A steel tube about eighteen inches in diameter and about as tall, it had a conical metal bottom resembling a wine bottle. Its top tapered like a funnel to a narrow opening, much like the Tin Man's hat in *The Wizard of Oz,* but not as folksy and inviting. It looked to me like one cone shaped-charge had eaten another cone-shaped charge on a high dose of steroids. A metal disc with a metal pipe about one inch in diameter and six inches long protruded from its side. Two wires came out of the top of the pipe. It was obvious this disc screwed to the top of the container to seal the narrow mouth.

The bomb techs put the device on the ground, and my partner and I approached it. We looked into the interior of the device and looked at each other in disbelief. The device was still about a quarter filled with an unknown explosive that I feared could be the infamous "Mother of Satan": Triacetone Triperoxide (TATP).

There was no way I would carry what could be potentially pounds of TATP in a vehicle with us. For all I knew, the bomb could have a fuzing system buried in the interior explosive. I looked back at the garden on the far side of the lot. A garden like that would need some caretaking. I asked one of the local folks if there was, by chance, a garden hose I could use to rinse out some of the "residue" in the device.

Our guests politely obliged. Rich and I took a small sample of the main charge for lab analysis. We then washed the approximately five to ten pounds of remaining explosive into the rose garden. Knowing that the main charge could also be one of myriad fertilizer-based explosives, I figured my pouring the bomb fill into the flower bed would either kill the plants long after we were gone or produce some of the best roses this side of the Israeli/Palestinian border. As long as the explosive was not in the car with me, I did not really care.

The empty device still contained a good thirty pounds of steel, and we heaved it into the back of our suburban. Along with it, we loaded the metal disc-and-pipe assembly. Our Palestinian colleagues told us they had made sure the device was safe. Of course, they also gave me a bomb with a good degree of main charge left inside. There were wires coming out of the pipe, and I was not really comfortable with that fact. I did not have the tools to examine this section of the bomb, and we had been exposed for too long in the open courtyard. My colleague and I decided to wrap this item in a bomb blanket I had brought with me and put the whole mess in the very back of the truck. The bomb blanket would stop any frag, in theory, which might be created from the explosion of this small piece.

We handed over to our Palestinian counterparts some packages of gloves, X-ray film, and other gear we had brought as a gesture of goodwill and said our goodbyes. Bomb stowed away, we went back into the guest house to see how the diplomatic niceties were proceeding. We sat back down and all eyes turned toward us, most likely because conversational material had thinned out while we were watering the garden. We noted that, indeed, we had secured the device. I thanked our counterparts for their assistance. The colonel in charge of the committee closed by launching into a speech about how peace-loving his people were, and how the typical Palestinian held no hatred for Americans.

We were in Palestine for the end of Ramadan, a period marked by fasting during the day broken by a large evening meal for celebration. The colonel looked at me, as I was obviously in charge of the bomb-recovery operation of the mission, and said he would take me personally in his car to Gaza City as a welcomed visitor at the evening meal, and he guaranteed I would see that "not a single stone" would be thrown at me.

A Belfast tour of graves and paintings on walls is one level of risk-taking. Agreeing to be driven into the heart of Palestinian territory for a dinner out is quite another. Luckily, the embassy people intervened and politely refused the offer on my behalf. To this day, part of me wishes circumstances would have allowed me to experience that feast.

As we crossed back into Israel, I asked our embassy escorts about the interior decorating of the guest house. As it turns out, the guest house sits between the Israeli border and an olive grove. From time to time, the Palestinians launched rockets into Israel from that olive grove. On the Israeli side of the border was a machine gun emplacement. To show their appreciation for the display of Palestinian aeronautical creativity, the Israelis sometimes opened up on the grove with machine gun fire. The guest house was simply a recipient of errant bullets meant for other targets. Rich leaned over and suggested that next time we visit we bring a couple of cans of spackle as a peace offering. I eased back and enjoyed being on the other side of the border.

The trip to pick up the bomb took us most of the afternoon. By the time we got back to the embassy it was getting dark. We still had one pressing problem. Rich and I both had suspicions about the metal disc with attached pipe and wiring. We asked our embassy colleagues to call the local Israeli bomb techs over with some gear. In particular, I wanted to get a decent X-ray of the bomb component before moving forward in any fashion.

X-rays are a common tool used in our field. In many cases, X-ray machines are used to take images of suspicious packages to determine if anything concerning is hiding inside. It was the perfect diagnostic tool to determine what we had on our hands. It took about an hour for the bomb squad to arrive. Showing them the pieces we brought back confirmed our fears that we had components of a fuzing system on our hands. Most bomb techs take apart all the components of a fuzing system before they throw it into the trunk of their car. My hope was that the Palestinians had the same sense of self-preservation, and the component we had was in an inert configuration.

The first X-ray of the device told me I might not be as lucky as I hoped. I could see the wires running into the top of the tube connected to a bulb positioned next to a dark mass. This is the hallmark of a detonator. The wires provide the energy to light up the light bulb. In turn, the bulb would heat up the big mass of dense supersensitive explosive that I was pretty sure was visible as the dark blob. The rest of the pipe seemed hollow.

Sitting inside an embassy office, we worked with the techs and their protective gear to slowly unscrew the small inner tube, which had the wires inserted into it from the main pipe. Mercifully, it was not glued into place. When I got it out, I could see the end stuck in the pipe had been covered over with a thick epoxy coating. It was a small sealed tube. It was also one of the biggest detonators I had ever seen.

An X-ray confirmed it as a full-up initiator. The mass of explosive contained within it was hyper-sensitive and capable of taking off my hand if something went wrong. Commercial detonators are fairly safe to handle if treated with respect. Detonators made by terrorists are not nearly as considerate of how you treat them. We gingerly wrapped up the detonator in one of our bomb blankets.

I looked down the pipe the detonator had been screwed into. It was packed with an off-white material that looked like damp, pressed flour. It looked like an intact booster system. The explosive in the bomb was a fairly robust mixture that needed a booster to set it off. In this bomb, the small detonator was placed next to this larger mass of explosive in the pipe to create a booster, or a larger charge that could set off the main charge.

So, to recap, we had basically carried a full-up explosive pipe bomb across the Israeli border wrapped in our bomb blanket. Had that part of the bomb gone off, it would have shredded the blanket and blown all the windows out of the SUV. If we were lucky, we would have only suffered minor thermal burns and hearing loss. If luck was out for a walk that day, a piece of the pipe frag could have created some serious injuries.

So what's the take-home lesson? Beware Palestinians bearing bombs. No bomb tech in any part of the western world would have ever kept the bomb component in the condition it was passed on to us. Every portion of a fuzing system needs to be separated, and any sub-components that contain explosives need to be put in a container that can withstand a blast. Those are standard operating procedures when handling IEDs. With this experience, however, it became obvious that not all places adhered to those standards.

With the initiator separated from the booster explosive, I felt a wave of relief come over me. Suddenly, I remembered that I had to take this cargo back with me on an FBI Gulfstream that cost millions of dollars. Even though I had multiple bomb blankets, and, intellectually, I knew that the most dangerous item was the detonator and it could do nothing in the packaging I had,

I couldn't help imagining the detonator going off somewhere midway across the Atlantic and scaring the shit out of the flight crew. Even if the worst it did was scratch the paint in such an event, I could see no way it would not lead to a rapid career death spiral. Needless to say, the ride home was the longest eleven-hour plane trip of my life, but thankfully we got home without incident.

The initial evidence that rolled in from the October attack on the US Embassy convoy didn't offer a lot to work from. In all, I had some wire, batteries, tape, and a switch from the fuzing system. All I had from the bomb itself were about a dozen metal fragments. From the fuzing system, there was a chance that fingerprints or hairs might have come along for the ride. Those items were sent to the relevant disciplines to probe that possibility. The fragments were sent for chemical analysis in the hopes that trace residue from the main explosive charge had survived and could be detected. This also assumed that the people who picked up the evidence didn't contaminate it themselves during collection. Our people are excellent at taking precautions against contamination; other less developed countries cannot be counted on to apply the same rigor.

The fuzing system was pretty basic. Batteries, a switch, and about a hundred feet of wire all pointed to command detonation. What the bomb consisted of was another matter. I did not have enough fragments left to make any guesses as to how it was assembled. But the bomb we picked up in Gaza was the lucky break I needed.

In these cases, my primary job is to determine what the bomb was made of and how the parts were assembled. Following that, I look for anything that can link the bomb to a suspect or group.

In some cases, I might compare two or more bombs to see if they share common attributes. The full-up bomb we recovered allowed me to do two of those things.

In the fragments I had recovered from the first bomb, some very unique sections of metal existed. One section was a metal ring I referred to as a flange. The fully recovered device had a flange of the same size and shape with the same number of holes to allow connection via bolts—all good signs that the two devices were related. Examining the full device taught me that this flange was basically the seal for the bomb's lid and was used to house the detonator and booster.

I also had some thinner sections of metal that looked like they had curved impressions stamped into them. The full device had an inverted metal cone attached to its bottom. This cone had concentric rings, like those in a tree, which covered its exterior. I was now able to make a pretty good guess as to the purpose those three or four mystery pieces of stamped metal served in the exploded device.

When metal containers filled with explosive blow up, they can fragment into smaller pieces, like a hand grenade, or they can create larger strips of metal the length of the container walls. A couple of the long strips of metal I had just happened to be the same length as the height of my intact metal milk-can bomb.

Based on the flange, the stamped rings in the cone, and the strips of metal matching the drum height, I had pretty high confidence the detonated bomb was a sister to the intact one. Without the intact bomb, I would still be guessing how the one used in the attack was constructed.

Since I had so few fragments to examine, each was fairly valuable in my assessment of the device. On top of that, the origin of the bomb fragments had to be taken as a matter of faith. Our team was handed the remnants of a bomb by the Palestinians who were able to get into the scene. I knew I had pieces of *a* bomb;

there was no way of being certain I had pieces of *the* bomb used for the attack. This might sound cynical, but if the case ever went to trial, it would be crucial to trace the bomb fragments to the attack without any question. In this regard, I also got lucky.

The SUV that was blown apart was recovered and sent to Germany for analysis. Why Germany? There is a strong US military presence in Germany, and transport in and out of that area is fairly easy. When the German crew processed the vehicle, it came up with a piece of bomb frag that had gotten lodged in the mangled interior.

In a normal case, I would have put that frag with the others and not thought twice about it. However, I wanted to be 100 percent sure all my evidence came from the device. Since the piece from the vehicle obviously came from the bomb used in the attack, I knew its origin without any doubt. The vehicle had to be torn apart to recover it, so it couldn't have been planted. If I could associate that piece with the bomb frags given to us by the Palestinians, I could have more surety that nothing underhanded had occurred. This is where metallurgy came to the rescue.

When steel is made, iron and carbon are mixed with several other metals to form the final product. Along the way, very small quantities of contaminating metal get added into the mix. In all, there can be dozens of contaminants in varying amounts. These other metals can be present at single-percent values, all the way down to very small fractions of a percent. The important thing is that the ratios are unique and hard to reproduce by chance (if not impossible). Our metallurgist analyzed both the metal frag recovered in Germany and the fragments handed to us by the Palestinians. To my relief, there was a match between the German frag and those provided to us by our partners. That result closed the door on any paranoia that might have crept in from skeptical investigators on any side.

A truly amazing result would have been for the metal from the Gaza attack to have matched the metal from the intact device we recovered. Fate was not that kind. At least the intact device had given me the critical insight I needed to understand how the device used to kill my countrymen was assembled.

Some gifts keep on giving. About a month after I returned from Gaza, I had one specialty examination on the recovered detonator to conduct. I had removed the booster charge from the pipe and found it to be a fairly robust and safe explosive. Analysis had also been completed on the sample of the main charge we returned with. However, the detonator still had to be examined. By design, detonators contain the most dangerous charge in the entire bomb. There may not be a lot of this material, but what there is does not take kindly to any external agitation. I had to take the detonator apart and remove that material. It is hard to do that without creating some degree of animus between you and the explosive.

Luckily, our lab shares an explosive demolition range with the United States Marine Corps. On base is a full explosive ordinance disposal unit. We have a long history of working closely with the Marines; they have been a fabulous asset to us over the years. Marines are unique in that they are the one branch of service that is allowed to take apart and analyze bomb components. On our shared range, the Marines had an entire building designed to remotely disassemble ordnance. My device contained a couple of grams of explosive filler. The Marines' building could withstand one hundred times what I had.

In the dedicated building, an item was placed in the room and a drill placed over a selected spot. Once the item was positioned, an operator stepped into the control room and, via remote control, began drilling, cutting, pulling, sawing, and a whole assortment of other destructive operations. The goal was to rapidly disassemble the item. All of this was done behind a thick bomb-resistant shield. If something went wrong, the operator was shielded. The event might alter the state of one's underwear, but it would not alter his life.

I made arrangements to take the detonator out to our range and meet with the Marines at their building. In theory, the operation was about as simple as it could get. I had a small tube with some hardened epoxy on its end. I needed to put it in a vise, position a drill bit to bore through the epoxy seal, and then pour out the inner powder. I had previously conducted this operation on dozens of commercial detonators. This was the biggest sample I had tackled, however, and I wanted the Marines there, as they were experts on their equipment.

Ranges are very busy places. Many groups need range time to conduct operations, testing, and research. The day we were slated to take apart my detonator, another group was testing a new tool for taking apart large fifty-five-gallon drums of explosive often seen in vehicle bombs. The large explosive test was scheduled to take place downrange from where I was working, and the charges being utilized were only about ten pounds. Mind you, ten pounds can tear a car to pieces. However, placed in the open a few football fields away, it will only make a loud crack.

The test was actually a continuation of a series of tests that had been conducted with this new tool. The researchers were set to shoot into fifty-five-gallon drums filled with ANFO. In all the previous tests conducted, the tool had knocked apart the drum and scattered the ANFO. In this final test, the researchers put some dynamite in the drum to simulate a more realistic charge. To save time, the test crew put out three drums of ANFO on the range. This allowed them to set up three tools and shoot three shots back-to-back.

There is something to be said about range limits. Every range has a maximum amount of explosive allowed to be detonated. This amount is usually dictated by the proximity of the nearest neighbors. In New Mexico, our ranges closer to town might only be allowed to shoot ten pounds. Those on the far side of the mountain could easily shoot five thousand pounds. On this particular range, the maximum allowed charge size was fifty pounds. As this test used about ten pounds of material, it was well below the allowed explosive weight.

In an ideal world, the tool would have gone off. Ten pounds of high explosive would have torn apart the fifty-five-gallon drum and scattered ANFO and dynamite across the ground to be picked up and disposed of at a later time. In a less-than-ideal world the ten pounds of high explosive in the tool would have initiated some portion of the much more sensitive dynamite explosive in the drum and a much louder explosion would have ensued. In a world where things approached "this shouldn't happen," a portion of the ANFO in the drum would also initiate and a mass slightly over the fifty-pound limit would detonate. The world I occupied that day was "you're about to be screwed."

The ten pounds of explosive in the tool slammed enough energy into the dynamite to set *all* of it off. This, in turn, created just the right amount of motivation to set off the entire four hundred pounds of ANFO contained in the targeted fifty-five-gallon drum. That first fifty-five-gallon drum then set off high-speed fragments, which reached the other two fifty-five-gallon drums of ANFO and initiated them, adding another eight hundred pounds of explosive to the party. At this very moment, I was leaning over the controls of a drill bit starting to drill into a Palestinian detonator. A massive shock wave from twenty-five times the maximum allowed amount of explosive on the range smashed into the bunker and rattled us like a fly between cymbals.

I honestly believe my heart stopped for a second. In the decade I had been detonating explosives, I had never been that close to such a massive charge. The battle-hardened Marine who was in the bunker with me fell back into the wall and slid down onto a crate. In shock, he yelled out, "Oh, shit!" on repeat. I thought he was going to have a heart attack. When my ears stopped ringing, I ran outside.

"What the fuck was that all about?" I shouted at the safety officer. I was livid, but the safety officer was more so. I backed down to let him go scream at the test organizers and went back into the bunker to complete the job I had started. When my hands stopped shaking, I easily drilled into the detonator and retrieved the powder to take back for chemical analysis.

In real-life forensics, cases don't open and shut within a sixty-minute window. Instead, fifteen years can pass with no big breaks, no breakthrough moments, and, sadly, no justice. I've done a good number of interviews over the years. You get used to hearing the stock questions: What was your most interesting case? Your most memorable? What case are you proudest of?

I can't recall one time ever being asked which case I regretted the most, so I suppose I'll pose the question here and answer it. Gaza is the case that possesses the most regret for me. Three Americans gave their lives to protect other Americans on a humanitarian mission. I wish I could have played a part in the story where their attackers were apprehended and brought to justice. Sadly, this was not the case.

Shortly after I recovered the bomb and did my analysis, we had another meeting with the Palestinians. I returned to Israel on February 5, 2004, to meet with the same Palestinian committee who had greeted us in Gaza. This time, I brought FBI forensic reports with me to show the good faith of the US Government. In return, we received assurances that the investigation in the Gaza Strip was ongoing, but no real suspects were ever developed.

Now that it was months after the initial detonation, I had a much better understanding of the bomb's construction. Through interpreters at the meeting, I asked my counterparts if the construction characteristics I was now able to attribute to the bomb used in the attack were indicative of any particular terrorist group. Perhaps knowing the bomb's construction would at least narrow down the perpetrators to one group.

My bomb contained a fairly unique attribute—the flange—and looked to be something that was custom made in a machine shop for this purpose. Surely, a group going to the trouble to fashion one such flange would make multiple copies and use them in multiple devices. Did my counterparts know about groups in Gaza that used such components? It seems there is no word in Arabic for flange. So, I drew pictures and tried to explain the purpose it served. In return, I was told there were two main types of bombs in Gaza, the teacup and the saucer, which meant nothing to me. To help, the Palestinians drew pictures of teacups and saucers. My bomb did not resemble dinnerware at all. Not to be dissuaded, I asked if either teacup or saucer might perhaps utilize a flange. This line of inquiry did not turn out to be productive at all. This exchange went on for about fifteen minutes before we abandoned it.

Yasser Arafat died November 11, 2004, slightly more than a year after the brutal attack. On January 9, 2005, Palestinians held a presidential election—the first since 1996—which took place in the West Bank and Gaza Strip. PLO chairman Mahmoud Abbas was elected to a four-year term as the new president of the Palestinian Authority. Oddly enough, the entire committee Arafat had appointed to assist in our investigation vanished. In its place, Abbas appointed another crew of people to lend us assistance. I met with this new crew just one time. On July 7, 2005, I traveled deep into the West Bank to a police station on the outskirts of Ramallah to revisit all the old briefings I had given the old committee, and to see if any more insight was forthcoming. The new team noted that they had indications that a well-known Palestinian terrorist group was behind the attack. That information was not exactly a surprise to any of us. That was all they would say about the matter.

On January 25, 2006, elections for the legislature of the Palestinian National Authority (PNA) were held. The result was a victory for Hamas, who won 74 of the 132 seats. With Hamas controlling the legislature, the last hopes of cooperation with the Palestinians on the investigation were lost and the case was closed.

Not every story has a happy ending. We're taught that a good story sets up tension and works characters toward a resolution. Some resolutions are victories for the heroes, others are tragedies. In real life, however, resolution sometimes seems to miss its cue and remains sitting in the stage wings.

CHAPTER 11

Rising to the Challenge

Bombers' motives and means are as old as the first black powder blast. Our ability to track down bombers has become more sophisticated, of course. But so have their means to "track down" each other.

We know that, historically, bombers attack out of anger over local government events that directly affect them. Labor disputes in the US, which began all the way back in 1910 with the *Los Angeles Times* bombings, are a prime example.[27]

Sixteen sticks of dynamite attached to a wind-up clock were used to destroy the *Los Angeles Times* building and kill twenty-one people. The building belonged to a rich industrialist, Harrison Gray Otis, who had formed a consortium of business owners in Los Angeles whose sole purpose was to keep unions from organizing. Otis became a symbol of the entire anti-union movement in the United States. He used the newspaper he

[27] It's worth noting that the earliest known Labor Union bombing of national significance dates back even further—1886 in Chicago.

owned to constantly beat the drum against those who would organize labor.

Out of this struggle was born the most prolific serial bomber of all time, Ortie McManigal. Paid by the leaders of the International Association of Bridge & Structural Iron Workers (IABSIW), McManigal personally deployed approximately one hundred bombs over his "career."

As information and current events became easier to disseminate more broadly via television and newspapers, it wasn't only those personally impacted by events who became potential bombers, but also anyone who read the news and took any particular offense personally.

Now, those in search of rage can simply open up the social media app of their choice and become inspired to be a soldier for a cause to which they have a tenuous, if any, connection. To say the internet has caused a revolution in that regard is to say the ocean is damp. As the Arab Spring—revolutions in the early 2010s fueled by social media platforms Twitter and Facebook—showed, it is now possible for just a few individuals to convey information to hundreds, thousands, or millions in the snap of an app. This has enabled change at a staggering rate. This fact has not been lost on terrorist organizations, as the headline pulled from the national news surrounding the Boston Marathon bombing makes abundantly clear: "Boston Marathon Bombing Suspect Dzhokhar Tsarnaev's Secret Online Life."

That was the headline from ABC News, on March 10, 2015, as the Western world woke up to the realization that violent jihadists are not only in faraway lands. They live among us.

On April 15, 2013, at 2:49 PM, two homemade devices, created by brothers Tamerlan and Dzhokhar Tsarnaev, detonated 12

seconds and 210 yards apart from each other near the finish line of the Boston Marathon. Three people perished and hundreds more were injured in the hopes of achieving, in Dzhokhar's words via his secret Twitter account, "victory over kfur [nonbelievers]."

Like most everyone else in the world, I tend to get the first indication of a bombing from reports trickling in from news outlets. Our unit gathered around the TV as clips captured during the explosions played in a continuous loop. Even over the TV, it was obvious that the explosive utilized was not military in nature. A chemist can tell a great deal by the visual indicators following an explosion. Black smoke shows an explosive lacking the oxygen necessary to consume all the carbon in the surrounding soup of atoms. Military explosives tend to be over-fueled and produce a dark-colored smoke. White smoke designates a family of explosives with much better balance between fuel and oxygen. The video clearly showed a light-colored smoke.

The amount of fireball, presence of a white flash, and thickness of smoke all provided clues, as well. Pretty quickly, we ruled out certain types of materials. But camera exposure is a tricky thing, and amateur video lacks rigorous color balance, so we were careful to not pass initial conjectures to anyone outside our unit.

Invariably, the next question was, "How big was the bomb?" In an episode of *CSI* featuring a bomb blast, one of the forensic "experts" answered this very question with this insightful evaluation: "It's not the size of the bomb, it's the overpressure in the air around it."

The first time I heard this response, I stopped the DVR and went back to make sure I was not imagining the reply. After confirming that I had, indeed, not misinterpreted the evaluation, I turned off the TV and took a shot from my tequila collection.

"It's not the size of the bomb, it's the overpressure in the air around it," is akin to stating, "It's not the size of the storm, it's how much rain it drops on your head." Small storms don't drop a metric shit-ton of precipitation. Big storms don't sprinkle. In fact, the bigger the bomb, the more explosive it contains and the more overpressure it produces. Overpressure comes from the bomb. It doesn't magically manifest itself in the air in a random fashion whenever a bomb happens to be in the area.

So how big were the bombs used in the Boston attack?

It was hard to make an assessment based solely on what we saw on TV. There are some things that can be used to estimate a charge size. Not only is the color of smoke a useful indicator, the amount of smoke can also be an informative factor. The more explosive, the more smoke. Some really good video clips showed the explosion and smoke cloud. From those alone, it was possible to say the bomb was more than a pound and less than twenty-five pounds. Those are the extremes. But to truly estimate charge size, a much more detailed analysis of the scene is needed, and, even then, estimating charge size can be tricky. Later, forensic clues allowed us to make a much better estimate of just how much explosive was packed in the IEDs the Tsarnaevs created.

The Monday Marathon bombing sparked four days of tension, terror, and frantic investigation. One thing to always bear in mind is that no one assumes a bombing event is an isolated incident. When the Marathon bombs went off, the first assumption was that there might be more to follow. This could have come in the form of secondary devices left behind to target first responders trying to lend aid to the injured and investigate the scene, or it could have manifested in yet more bombings at any number of unanticipated targets. There was no way to know if the two devices were the finale or the prelude to a larger bombing campaign. Thus began round-the-clock crisis operations and criminal investigation. There were three agonizing days of this uncertainty before the FBI had suspects and photos of them, which could be released to the public.

Once the photos of the Tsarnaev brothers were released, a torrent of chaos followed. The brothers—realizing that the combined citizenry of the entire Northeast was focused on finding them—gathered up their leftover bomb-making paraphernalia and attempted to escape, hoping to pull off one more attack. Twenty-seven hours of mayhem ensued.

Armed with one pistol, the brothers attempted to get another firearm by ambushing a twenty-six-year-old MIT police officer in his car and shooting him in the head. A complex locking mechanism on the officer's holster thwarted their attempts to get his service firearm. From there, the brothers carjacked an SUV and held its owner hostage for about thirty minutes before he escaped at a gas station. They identified themselves as the wanted bombers to the carjacking victim, who immediately contacted authorities upon his escape. What followed next was a whirlwind of desperation on both sides.

Police began a manhunt for the SUV and spotted it in Watertown, where a firefight ensued between the bombers and the police. During this firefight, the brothers are reported to have thrown "hand grenades" at the pursuing officers. The older brother, Tamerlan, bolted from the car, engaging in a gunfight with police as he ran at them. He was shot. His younger brother, Dzhokhar, attempted to ram the officers who were engaging in the gunfight with his brother. In the process, he ran over his brother with the SUV, ostensibly finishing off the job started by the police. But Dzhokhar escaped.

The people in Watertown were ordered to shelter in place, an order that lasted for approximately twelve hours on Friday (the fourth day after the bombing). Eventually, with Dzhokhar not found, the order was lifted. One Watertown man went outside to smoke a celebratory cigarette and noticed his boat cover had been moved. He notified the police immediately, believing someone was hiding in it.

At this point, the tensions and emotions of local police were heightened. Two horrific bombs had gone off four days earlier. Since then, the two identified suspects had brutally killed a police officer, engaged police in a shootout, and thrown bombs. Now the remaining survivor was trapped in a boat in someone's backyard. As you can imagine, lots of police responded to the scene. Like piranhas on a bleeding cow carcass, they schooled into the area.

I am sure someone tried to come up with a systematic plan to extricate the bomber from the boat, one designed to minimalize potential gunplay. I don't know if anyone to this day knows who fired the first shot. But a torrent of rapidly accelerated lead followed.

There is a scene in one of my favorite Clint Eastwood movies, *The Gauntlet,* where police open up fire on the house where they believe Eastwood and his fugitive protectee are held up. By the time shooting ceases, the house collapses under the structural damage imparted by all the bullet holes. Pictures of the boat Dzhokhar was hiding in always remind me of that scene. How he was not killed in the barrage of projectiles that worked their way through the boat is a mystery. How any of the residents in houses anywhere in the vicinity were not winged by an errant bullet is equally unfathomable. But Dzhokhar survived and was taken into custody.

So how did we figure out who planted the bombs and work out how they were assembled? This is where it takes a village, as they say.

The first bomb went off at 2:49 PM near the marathon's finish line. Within thirteen seconds, the second bomb exploded a block back up the street. When the final tally was taken, three were dead and approximately 264 injured. Among the fatalities was an eight-year-old boy. The bombs were placed low on the ground so that, when they exploded, a wave of pressure rippled outwards at calf level. Of the injuries sustained, fifteen people required single amputations and two lost both of their legs. Pictures from the scene show the sidewalks literally red with blood.

Working any bombing scene is complex. You must initiate multiple complex missions simultaneously. People need to be evacuated from the area. Injured need to be tended to. Finally, evidence needs to be preserved while rescue operations are underway and eventually collected. The Boston Marathon bombing offered some unique challenges.

Evacuation and rescue were made exponentially more complicated by the fact that so many people were congregated in the area at the time of the blast. Many people had backpacks and purses. In addition, folks near the finish line carried gym bags with clothing for runners to change into after the race. When the explosions occurred, fight-or-flight reflexes took over, and the crowds scattered. Many people left behind the bags, purses, and backpacks they had brought with them.

In any bombing, there is always a chance for a secondary device. Eric Rudolph, the Centennial Park bomber, brought this lesson home in January 1997 when he bombed an office building that housed an abortion provider. Rudolph set a time bomb to go off behind the building. Little to no damage was imparted, and responders got to the scene quickly. What the police did not realize was that the first bomb was merely bait. Rudolph had set a second time bomb secreted approximately one hundred feet away near a large trash bin in another corner of the parking lot, a location where police would logically stage a command post to investigate the first blast. The bomb, comprised of dynamite and hundreds of nails, was set to detonate an hour after the first blast. But for an ironic turn of events, Rudolph's second bomb would have sent a lethal barrage of shrapnel into the assembled responders.

One of the workers at the building had recently purchased a brand-new BMW convertible. The new car was just out of the fragment zone from the first blast. As police set up crime scene tape and got ready to process the scene, the man asked if he could move his car to the front of the building. He was given permission to do just that. In an effort to protect his new car, he parked it away from the main lot on the other side of the police and fire vehicles. He found the perfect place right in front of a trash bin. When the second bomb went off, his brand-new BMW served as an ideal barrier to absorb the air blast and hundreds of nail fragments that were aimed at the responding officers and firefighters. His car was demolished, but its heroic sacrifice saved many lives. The bomb is caught on the video of a reporter setting up for a shot. The camera captures the explosion and the blast wave puffing up the reporter's hair as it passes harmlessly by.

Back in Boston, responders were faced with a scene where dozens and dozens of potential secondary devices were scattered across two crime scenes and interspersed throughout all the rescue efforts being conducted. It took hours to ensure the scene was safe for evidence collection. Many bomb techs heroically went from bag to bag checking them to ensure a ticking trap was not lurking amid the carnage.

As if the chaos at the scene was not enough, Murphy, law in hand, decided it would be a good time to ring the doorbell. Within five minutes of the Marathon bombs exploding, reports came in of an explosion at the John F. Kennedy Library about three and a half miles away. In times of panic, communications become chaotic. The purported explosion later turned out to be an unrelated mechanical fire. However, the conspiracies raged on for almost two and a half hours, diverting resources and attention away from the actual crisis point at the finish line.

I cannot recall another event in the history of US bombings where two bombs detonated within such close proximity in both space and time. The circumstances of the attack created two distinct crime scenes. The physics of explosions created two scenes with overlapping evidence. Only one block separated the two devices, and both had ample energy to toss pieces into each other's backyard. Processing the scenes required good coordination and expert analysis.

On any given day you can ask me how I feel about working for the FBI. There are days when I face the typical government bureaucratic stupidities where my response might lack flattering prose. However, I have worked long enough to see the FBI respond to numerous crises. On those days I stand in awe of the juggernaut we transform into. I almost imagine some powerful leader in the lofty halls hearing of the latest attack and reaching over to the red emergency phone to utter a simple three-word phrase: "Release the Kraken."

The FBI has teams dedicated to the collection, documentation, and preservation of evidence. As previously introduced, these are our ERTs (Evidence Response Teams), made up of eight specialized individuals who focus on this mission. In Boston, the FBI sent multiple ERTs to each site. And the FBI deployed dozens of bomb techs and a team of Explosive Device and Chemistry examiners in fully outfitted deployment trucks. All of these assets converged on Boston while responders on the ground were tending to the dead and wounded.

ERTs, employing bomb techs as expert advisors, worked the two bomb scenes. Bombings are a very specialized area. We learned early on that such scenes require bomb techs to provide advice on what items to collect and what items to leave behind. At one scene years earlier, an ERT leader wanted to submit an air conditioning unit that fell from an embassy window following a bombing. The ERT leader noted that he felt the unit exhibited blast damage. Somehow the fact that the entire façade of the building smashed by the force of a car bomb also exhibited blast damage, and would serve equal value if packed up and sent to the lab, escaped this individual.

So, what was I doing at this time? When the Marathon bombings occurred, I was a few years into my tenure as the FBI's Chief Explosives Scientist. Forensics was not my day-to-day focus anymore. The FBI had a very talented cadre of device examiners, and my services were not needed to address such a foundational case. I had two jobs. The first was to do anything I could to free up time for the people doing the actual bench work, which included volunteering to pick up the evidence collected by ERTs delivered to a facility near my house and drive it to the lab at around one o'clock every morning. There, teams worked 24/7 to process specimens collected and conduct forensic evaluations. My serving as evidence chauffeur allowed an already strained team to have one less logistical headache.

After I delivered the evidence, I was updated on the teams' observations and what exams were being conducted. Then, as part of my second job, I kept upper management informed on the forensic results particular to the bombs. By stepping into this role, the forensic examiner on the case—an ex-Navy SEAL of few words—could focus on the critical task of putting back together the bombs without constantly being distracted by a barrage of senior-management questions. We refer to this aspect of the job as "feeding the beast." The Beast is imbued with both an insatiable appetite and an uncanny ability to interject requests for information at the time when bench work is the most demanding.

As much as it might be cathartic to answer the question, "What is taking so long to do the analysis?" with a retort of, "The delay results from interruptions by your constant phone calls, emails, texts, and casual in-person visits," such honest communication is not career enhancing. My job was to lift some of that burden off the folks doing the real work.

As bomb parts were collected and analyzed, a picture of the devices used during the Boston Marathon bombings began to emerge very quickly. One of the first things discovered were fragments of six-liter pressure cookers. In some explosive devices, a container like a pressure cooker can be blasted into bits smaller than a dime. Luckily, the explosive charge selected by the Tsarnaev brothers lacked that kind of energy. The bottoms of both pressure cookers remained intact. This was critical in that both were stamped with the manufacturer and model.

Every aspect of the evidence provides potentially valuable insight that might generate leads for investigators. In this case, the brand of the pressure cooker found was a well-respected one sold at only select high-end stores. This was a lucky break. Had the pressure cookers been sold at every Walmart across the country, there would have been many more leads to run down.

Once the brand and model of cooker was known, the FBI tracked down all of the stores in the geographic vicinity that sold them. As there were two pressure cookers, the FBI hoped that there might be some record of an individual purchasing two of these items. Maybe an alert store clerk would remember a customer buying two of a product that rarely sells one at a time. People who pay attention to odd things occurring in their place of business have assisted the FBI in the past. No potential lead can be ignored. With the identity of the bomber(s) unknown at the time, these were the types of leads that dozens of agents were running while the rest of the evidence was analyzed.

With agents scattered across the Northeast tracking down pressure cookers, the next item to be identified was what is referred to as the "concealment container." Bombs need certain things to function (like explosives). They also have attributes that might not be necessary for function but enable a broader purpose. One such attribute is a means to conceal their presence. In the case of the Marathon bombs, the devices were hidden in backpacks, which are most definitely not designed to withstand close proximity to an explosive charge. Most of what was found were small fragments of cloth, zippers, and other closure hardware. But a tenacious surviving component provided the next forensic clue.

Zipper pulls for both backpacks were located at the scene. As fairly robust stamped metal objects, zipper pulls tend to survive bomb blasts. Manufacturers also tend to use the surface area of zipper pulls to stamp their logos. Fragments of zipper pulls and some labeling tags recovered allowed examiners to determine the manufacturer and model of the backpack. This information was again fed to field agents who investigated how the backpacks were distributed. Detective work turned up that one backpack had gone out of production years ago, and the other was sold at a wide variety of big-box stores. Neither evidence item provided clues as to where they were purchased, or who might have purchased them.

The chemical residue left behind by the bombs was examined within the first few days. Sometimes investigators can get lucky and bombs may be poorly designed, leaving behind unconsumed material. But both Marathon bombs fully functioned, leaving behind no bulk explosive. In this case, trace residue was the only hope to identify the main explosive charge.

Chemists identified a variety of trace chemicals indicative of an energetic material. Since a bomb obviously exploded, the presence of explosive was not too enlightening. But the assortment of chemicals that can be used to make an explosive is fairly restrictive. For this reason, many different explosives employ similar chemicals.

To understand the power of the Marathon bombs some explanation of explosive categories is needed. All explosives are divided into two buckets: high explosives and low explosives. Both high and low explosives undergo rapid combustion reactions to produce large amounts of hot gases under high pressures. The only difference between the two is just how fast the combustion reactions occur.

The slowest-burning explosives are called low explosives. Even this "slow" burn can occur faster than human perception can take in. Most people have some passing experience with low explosives. The two families of low explosives that most people experience in everyday life are pyrotechnics and propellants (such as black powder).

High explosives are the fastest reacting of all explosives. They are the most powerful, destructive, and deadly. Hollywood has made high explosives such as TNT, dynamite, and ANFO household names. Bombers strive to get their hands on high explosives but will settle for low explosives if that is all they can acquire.

The residue left behind after the bombings and the sizable fragments were consistent with what is expected from a low explosive. But there was no way of telling *which* low explosive might have been present. Best guesses put it at either a black powder substitute, like Pyrodex, or a pyrotechnic filler. Investigators would have to pursue both avenues in search of the perpetrators.

Some bomb fragments were of very limited value. Early on in the analysis, copious amounts of added fragmentation were analyzed. The bombs were assembled with the intent to maximize injury or death. Additionally, cardboard sheets with pounds of additional fragments (BBs) adhered to them were packed around the interior of the pressure cookers. Hundreds of BBs were recovered, the kind sold at every Walmart and Target store across the country. Still, agents did what they could to seek out purchases of BBs in the Northeast.

Battery fragments also came into the lab. Some batteries have more of a story to tell than others. The garden-variety generic AA battery provides limited insight. However, the batteries recovered at the scene were of a very exact nature and specialty manufacturer. These batteries were sold for the express purpose of powering toy cars. From there, the lab was able to identify the make of the toy car used. Fragments of the other power pack, coupled with sections of a circuit board, allowed electronics experts to identify the toy car used in the second bomb. Two different toy cars created a richer pool of potential leads—and these were also pursued.

As agents tracked down sales of all of the bomb pieces, biometric examinations were conducted on all items submitted to the lab. Every item was exploited for fingerprints, DNA, hairs, fibers, and any other clue that might be of value. With an unknown number of bombers still on the loose, any investigative lead with the possibility of providing an identification was invaluable. For example, the presence of two different DNA signatures might indicate at least two bombers involved. Fingerprints were developed, but they would not be instrumental in identifying the bombers. Instead, the lucky break came as a result of Herculean video analysis.

Bombs may have a lot of ass, but they lack legs. What I mean by that is that the two explosive devices did not take themselves to the marathon finish line. It stood to reason that someone may have seen something that could provide investigators some sort of lead on the bomber(s). The location of the attacks favored the investigators. Numerous restaurants and storefronts lined the street leading up to the finish line. There was a chance something may have been caught on one of their security cameras. In addition, hundreds of people crowded the finish line waiting for loved ones to proudly cross the line, many with cell phones in hand recording.

The FBI has a group of folks who specialize in video analytics as well as teams that specialize in creating massive databases. Shortly after the bombings, the FBI set up a server where the public could upload pictures and videos that were taken at the marathon. Within ten minutes of the FBI announcing the server to the public, it almost crashed from the crush of data being uploaded to it by concerned citizens. In TV crime dramas, computer algorithms would parse through terabytes of data containing hundreds of thousands of images and videos searching for anomalies. In the real world, this is done by teams of people.

Analysts focused on videos near the two areas where the seats of explosions occurred. Eventually, they found a picture showing one of the backpacks on the ground right before the explosion. Now they had to find out who put it there.

One video clip turned the tide. A single uploaded video shows a crowd of people staring straight ahead at runners crossing the finish line. Suddenly, an explosion happens off screen and the crowd en masse pivots in the direction of the blast—all, that is, except one lone individual who turns the other direction and slowly walks away. The first bomber was identified.

The individual captured by that video clip was Dzhokhar Tsarnaev, known to investigators initially as "White Hat" for the white baseball cap he was sporting. Once investigators had a person of interest, all the video in the area was analyzed to find clips where he was present. He was quickly identified as having entered the area with a second individual, later identified as his older brother, Tamerlan, dressed in a black ball cap (he became known as "Black Hat").

Once these two suspects were identified, a series of events occurred that necessitated releasing their identity to the public.

I have been involved in many high-profile cases over the years. In every one, there was a constant stream of leaks to the press. Once I recall waiting for the UK to send some pictures of a fuzing system over to our unit for examination. We were on the phone with FBI counterparts in London trying to get the file expedited. While we were waiting, I looked up at the TV, where CNN was running quietly in the background, and told our rep overseas that there was no rush anymore because the news had just posted a picture of the item of interest.

I am all for telling the American public what they need to know to be vigilant and safe; however, in my experience, the media doesn't care about public safety as much as they care about scoring a salacious fact or graphic image to draw viewership. In the Marathon bombings, the press somehow got ahold of pictures of the brothers and was going to publish them, regardless of what the FBI wanted. This situation allowed law enforcement little time to develop the best investigative strategy and forced a public release of the pictures of Black Hat and White Hat. The net result was a cascade of actions where the brothers felt forced to act rapidly, leading to the previously described tragic murder of the MIT police officer, and the Watertown shoot-out with police. I am not directly blaming the media for the way things unfolded, but I will say I have never had any positive experiences with media during any bombing investigation. In many cases they have been more help to the bombers than my team.

A full description of how the bombs functioned can be found in the trial testimony, and accompanying news reports, provided by the ex-Navy SEAL forensic examiner that I spoke of earlier. News outlets vacillated between calling the bombs "relatively sophisticated" and "not sophisticated." In truth, they were a little of both. The main charge was culled from pyrotechnic display pieces. The bombers harvested a variety of explosives from these devices. In an air burst device, a lifting charge of black powder propels the exploding projectile into the air. Black powder is the most voluminous material in most aerial shells. Many mortars had been cannibalized by the bomb makers for their black powder. Exploding shells launched by mortars have a wide variety of explosives encased within them. These include a flash powder charge to produce the echoing report (or boom) and numerous small round pellets thrown outwards that produce the arcing colored trails. All of these were also harvested. Each device contained about fourteen pounds of such material.

In terms of sophistication, it does not take much ingenuity to take apart fireworks. Juveniles have been engaged in this activity for decades. It is hard to say we got lucky in the bombers' choice of explosive. However, things could have been much worse if they had used some of the other explosives we routinely see applied in explosive devices. Some of these explosives would have created much more damage and resulted in higher fatalities. In a glass-half-full view, the lack of sophistication in the main charge limited the damage the bombs could impart.

Although not of military-grade power, fireworks can obviously create a horrific bomb. As the main charge in the Marathon bombs became known, various questions emerged within the FBI about how future attacks with such a mixture could be avoided. The FBI's Weapons of Mass Destruction Directorate (WMDD) is tasked with outreach to industries and companies that carry chemicals that can be used to make bombs and poisons. I have worked closely with them for years on explosive precursor chemicals (EPCs). WMDD wanted to put out a bulletin to fireworks sellers regarding suspicious purchases to make them aware of the potential misuse of their product. As part of these bulletins, people were advised to be on the lookout for suspicious behaviors. In the first draft of the bulletin, this advice was paired with telling sellers to watch out for folks asking for the biggest firework available.

This recommendation, written by a well-meaning analyst I am sure, showed the challenge faced in outreach to the fireworks industry. Every single consumer that goes into a fireworks shop seeks out the biggest, loudest, and ballsiest exploding article they can afford. Treating someone looking for the biggest firework as suspicious is like zeroing in on all customers who Supersize their already jumbo-calorie, cardiac-crusher combo meal. Need has nothing to do with desire. On the other side of the purchase, fireworks stores typically put their most opulent, over the top, Rambo-inspired pyrotechnics in positions of high visibility. They want these things front and center with a large spotlight on them to entice customers to sink as much money as possible into the purchase. Warning fireworks sellers to be on the lookout for people who want huge quantities of oversize pyrotechnics is basically telling them to send in their whole customer list every day. This was not an easy vulnerability to address.

Fireworks do not require much energy to ignite them, which is why a cheap piece of burning fuse can supply enough heat to set off a firecracker. The ignition system utilized by the bombers consisted of a Christmas tree light bulb placed in the pyro filler. When energized, the filament provided enough heat to set off the explosive. This is also far from rocket science. Such an ignition source is one of the most common initiators we encounter in explosive devices. Even the pressure-cooker housing was far from innovative. Pressure cookers are a logical container for a low explosive like black powder and have drawn attention from bomb makers for years. Being metal, large, and able to hold a degree of pressure, they meet all the criteria for Bomb Making 101. Perhaps the only place where sophistication crept in was in the fuzing system of the devices.

Both devices were set off by fuzing systems created from toy car mechanisms. Remember, a bomb needs two things: an explosive and a way of setting that explosive off. In the case of the Marathon bombs, the explosive was set off by the Christmas tree bulb lit up by a modified toy car. Two different toy cars were utilized, one for each device, to ensure that there was no accidental cross talk between the cars' transmitters.

Both devices were initiated by pressing the power switch on the controllers. Again, this is not sophisticated either. The evening before one of my regular trips to FBI Headquarters to brief Director Mueller and other top members of the Bureau, I was told by the lab director that Mueller wanted to know all about toy car transmitters and their use in explosive devices. At the time, some debate existed about how sophisticated this tactic was. Some argued that such a fuzing system was evidence the brothers had received advanced bomb-making training.

I spent an entire evening on YouTube to see what was common knowledge.

What I found was a glut of videos providing instruction on how to turn toy car transmitters and receivers into pyrotechnic firing systems. These tutorials showed an easy way to modify a toy car to allow the viewer to set off fireworks at a distance. Notably, the vast majority of these how-to lessons were narrated by pre-pubescent boys. The next day, I disabused our executives of the notion that toy-car-based triggering systems were indicative of state-sponsored tradecraft. At best, they were part and parcel of the plethora of knowledge promulgated by the "Youths of America over the Interwebs." Mueller never smiled at any of my turns of phrases, but I know some of the other executives appreciated the levity.

One of the toy car transmitters was found in the vehicle that was recovered after the Watertown shoot out. Typically, these transmitters possess what is referred to as a pistol grip and look like a stun gun on steroids. It would be difficult to use the transmitters in "as sold" form without anyone in proximity noticing a massive pistol-shaped object. Therefore, the bombers disassembled the controllers so that the control board and battery pack were the only things left. Taped together, they easily fit in a coat pocket. That modification was slightly creative.

The bombs utilized in the Watertown attack were even more basic than those used at the marathon. When the brothers saw the news break with their images being broadcast across the world, they realized time was no longer on their side.

They gathered up all the bomb-making materials they still had in their possession and attempted to make a run for it. Evidence indicates that they were initially thinking of taking the bomb-making materials to New York City and setting off a device there. This potential plot never had time to coalesce.

One of the bombs utilized in Watertown was a smaller version of the Marathon bombs. During the altercation with police, the brothers threw a smaller pressure-cooker bomb out of the car. This was a four-liter pressure cooker equipped with a piece of burning hobby fuse. Also deployed against law enforcement were small explosive devices the press dubbed "hand grenades." In truth, these were just small pipe bombs. The devices consisted of small steel pipe elbows and couplets filled with the same pyrotechnic filler as their larger cousins, modified with a piece of hobby fuse running into their interiors. One unique aspect was that the pipes also had the same BBs seen in the Marathon devices glued to their interiors for additional fragmentation. Some of the devices exploded while others were recovered intact. Plenty of pictures exist on the internet for those curious about their appearance.

When going into a trial, the whole purpose of forensics is to bring scientific evidence into the courtroom that either further ties a subject to a crime or exonerates them. When the stakes are as high as they were in the Marathon bombings, which was a death penalty case, the science needs to be beyond question.

Donald Rumsfeld once said, "You go to war with the army you have, not the army you might want or wish to have at a later time." Much the same can be said of trials. You go to trial with the evidence you have. It is the job of each side to spin a narrative around the investigative facts and the scientific evidence in support of their contentions. My unit's focus was on the explosive evidence, but other forensic disciplines obviously played a key role.

The strongest way to tie someone into a crime is by what scientists refer to as biometrics. These are the biological signatures that point back to one subject above all others. Testimony at the Marathon bombings trial showed that Tamerlan's fingerprints were found on multiple pieces of evidence: remnants of the first explosive device, the lid of a pressure-cooker bomb thrown in the Watertown shootout, recovered suspected bomb-making materials (rolls of tape, a caulking gun, a soldering gun), and receipts for the toy cars from which the fuzing system was assembled. In addition, Tamerlan's prints were on the one transmitter recovered after the Watertown shootout.

The fingerprint evidence was used to support both prosecution and defense stories. It was obvious that Tamerlan could be tied into the bombings. Matching the old turn of phrase, his prints were all over the thing. Tamerlan, however, was dead. On trial was Dzhokhar. And his fingerprints were found on nothing associated solely with the two bombs. The defense could not contend that Dzhokhar had nothing to do with the bombings. Remember, there was video of him carrying one of the devices on his person to the scene.

Investigative evidence had him tied to the event. Forensic evidence failed to tie him to the bomb construction. The defense used this lack of forensic evidence to paint a picture of a younger brother who was led by the authority of his older sibling, an older sibling who was 100 percent responsible for making the bombs and coming up with the plot. To the average person, this may not be much of a distinction, as Dzhokhar did carry and place a bomb, but in a death penalty case such distinctions can make the difference between a life spent looking at concrete walls and one with a lethal injection curtailing it.

Explosive-related evidence was also used to link the bombers to the horrific crime. Post-blast residue found on the devices was already discussed. The chemist knew that it was consistent with a low explosive. Once the brothers were apprehended and their residences, vehicles, and other affiliated locations were searched, more evidence was sent in. ERTs searching such locations take with them specialized tools. One is a forensic vacuum. It works on the same principle as a home vacuum but collects minute quantities of materials it sucks up on sterile filter paper. The hope is that such vacuuming might sweep up small particles of unconsumed explosives. Tearing open fireworks is messy. Young men are not known as meticulous housekeepers, as any parent who has visited a child in a frat can testify. In the Marathon case, the vacuuming did yield valuable evidence.

In all, the explosive chemist analyzed about three hundred pieces of evidence. Vacuum samples sent in from the Tsarnaev family apartment in Cambridge contained grains of black powder. Residue consistent with the explosive utilized in the bombings was also recovered from the Honda CRV driven by Tamerlan. Latex gloves found in the car had residue recovered from their fingertips. Again, forensic evidence needs to be taken in context. There are legitimate uses for pyrotechnics. However, just the purchase of pyrotechnics will not result in the contamination of a car and house with black powder. Taking apart those pyrotechnics to harvest their explosive filler increases the odds of such contamination substantially. Although, just like in the case of the fingerprints recovered, the traces of explosive tied the older brother more strongly to the bombs. However, a piece of almost-lost evidence linked back to the surviving younger brother. This evidence was recovered from a backpack (and not one used to carry a Marathon bomb) owned by Dzhokhar.

Shortly after seeing their friend's visage splashed all over national news as a suspect in the bombing, three of Dzhokhar's pals went to his dorm room and removed a backpack filled with fireworks. Some of the fireworks appeared to have been manipulated and emptied of their gunpowder filler. In a state of panic, and imbued with badly placed loyalty and youthful stupidity, the three put the backpack and its contents in a garbage bag and tossed it into the dumpster behind their residence. Eventually, investigators uncovered this fact. This led to a two-day search at the New Bedford landfill, which miraculously turned up the discarded evidence.

Based on their role in the coverup, the three friends were sentenced to three, three and a half, and six years in prison, respectively.

As for the bombs themselves, only one toy car transmitter was recovered. The fact that this transmitter was found in the getaway car used by the brothers and the bombs' fuzing system used a modified toy car of the same brand might seem to be strong enough evidence to associate the brothers with the bomb. However, the goal of forensics is to make as strong a tie as science can create. Our electronics folks, who can read the secret codes deeply embedded in all things digital, determined that the transmitter recovered had been paired with the receiver in the bomb's circuit. One can argue that anyone could have a modified transmitter lying around in their car, although it would be a fairly lame argument. It is harder to argue that the modified transmitter found in your car just happened to be matched up to communicate with the circuit in a bomb that went off.

Other evidentiary items confirmed the Tsarnaevs were without a doubt associated with the bombs. Found at the Tsarnaevs' family apartment in Cambridge were nails, BBs, and pressure-cooker parts. Modifications had to be made to the pressure cookers to transform them into explosive devices. Such modifications left extra parts. Finding a gasket to a pressure cooker and the top of a pressure cooker lid without the rest of a pressure cooker again told a very convincing story to the jury during the trial. And a string of Christmas tree lights, with some of the bulbs cut out, was also recovered from the same location. The evidence found correlating back to the construction of the bombs was fairly overwhelming.

There was an interesting link between the Tsarnaevs' bombs and AQAP's *Inspire* magazine, specifically their flagship bomb-making article referenced at the start of this book: "How to Make a Bomb in the Kitchen of Your Mom" by the "The AQ Chef." While the bomb-making instructions included in this article were far from original, sophisticated, or creative, they were effective. Based on the premise that bomb-making recipes should be as universal as possible and require materials that can be acquired from anywhere in the world, the article outlined the production of a simple pipe bomb. It was from this article that the genesis of all the Tsarnaevs' IED construction ideas emerged.

The article focused on pipe bombs created from match heads and pipe elbows with simple cannon-fuse igniters. Placed side-by-side, the pipe-bomb elbows used during the shootout at Watertown are dead ringers for the pictures of the pipe bombs depicted in the article. In an odd twist, it was the sidebars in the *Inspire* article that created the road map that the brothers followed for bomb production. In a sidebar adjacent to the main article, advice is given to those wishing to build larger bombs. The writers recommend pressure cookers take the place of pipes. An explosive fill based on match heads is called out for in the article. But a side paragraph notes that those who want to build a big bomb may not want to harvest explosives from thousands of matches (as I had tried earlier). To those ambitious bomb builders, *Inspire* recommends using either black powder or firework powder to fill their bombs. Added fragmentation is discussed, as well, and is seen in the Marathon bombing devices. Electronic copies of *Inspire* were found on every computer seized during the search of the two brothers. It became obvious that this document played an instrumental role in their plot.

There are always questions that go unanswered in a case like the Marathon bombs. The two big ones were: where were the bombs made, and did the bombers do any testing beforehand? If a bomb factory existed, it was never found.

Where exactly the dozens and dozens of fireworks were taken apart and explosives blended remains a mystery. It also remains unclear if the bombers tested out any of their design before deploying the bombs. Testing a fuzing system is simple. All that's required is to see if pressing a button causes a light bulb to glow. Knowing that an explosive charge will perform as desired is another issue. As both bombs underperformed, it is doubtful that the brothers tested any large amount of explosive material. Instead, it seems they depended on the advice offered by the AQ Chef. Luckily, for once, there were not enough cooks in the kitchen.

As mentioned earlier, I did everything I could to take pressure off the forensic examiners working around the clock to put the pieces of the Marathon bombing back together. While specialists examined fragments of the bomb under microscopes and extracted residues for analysis, I joined Mark Whitworth, who had become the chief of the Explosives Unit, for twice-daily secret video teleconferences (SVTC) with the heads of the FBI. Sometimes Director Mueller himself would join with questions. The Beast is always hungry; in this case, it was voracious.

Information could not be supplied fast enough to satiate its appetite in the first three days following the bombing. Once the bombers were apprehended, the pressure alleviated slightly, but the possibility of undiscovered co-conspirators still needed to be addressed. Luckily, none were found.

After a major bombing occurs, the Explosives Unit eventually makes a reproduction of the bomb based on its analyses. Sometimes this is done to show the potential destructive power of the device. In the Marathon bombs, this demonstration was not necessary. Many videos and pictures, and a tragic number of dead and wounded, spoke loudly enough to those points. However, a mock device is always useful in educating non-bombing specialists on how the IED functions.

While forensic examiners were still busy plowing through evidence, some colleagues and I put together a couple of mock-ups of the devices.

Early on, after identification of the pressure cookers via their stamped markings, we purchased about a dozen each of the same model of pressure cooker for reference. So, I had plenty of spare cookers to modify. We also purchased about a dozen of the exact toy cars. With plenty of sacrificial cars and cookers at my disposal, it was easy to craft a working model (minus the stuff that goes boom) to show our executives.

The devices were not light. Part of both bombs were what came to be referred to as frag sleeves. As mentioned previously, the bombers had taken cardboard and affixed hundreds of copper-coated BBs to them. It was my job to reproduce these in our mock-ups. Early on, I had purchased huge boxes of donuts for the teams working nights (one of my more useful contributions). Looking for cardboard, I scavenged the lid from one of those boxes and cut a rectangle out of it. The rectangle was shaped so that it could be curved around and just fit into the pressure cooker as a sleeve. Next, I covered the cardboard with a layer of adhesive and layered BBs across the whole sheet. This was just what was recommended in *Inspire*. It was a time-consuming and messy process, but when the sheet dried, the BBs adhered to the cardboard and the sheet was stuffed in the pressure cooker.

BBs don't look like they weigh much. As individual metal spheres they weigh almost nothing. This changes when you get thousands of them together in one place. I ended up with about fifteen pounds of BBs by the end of my experiment. This, coupled with the pressure cooker, made for a hefty device. We rigged up the modified toy car so that when the transmitter was turned on a loud buzzer would sound from inside the pressure cooker. This simulated the fuzing system setting off the bomb.

For good measure, I also reproduced the pipe bomb elbows and couplers used by the bombers. Both the steel pipes had BBs coating their interiors. Once finished, they weighed a few pounds each. In fact, they weighed enough that I was fairly sure I would not be able to throw them far enough away to ensure frag would not travel back at me if they were real. The two pipe bombs and the BB-laden pressure cooker were all placed in a large backpack with the transmitter. When all combined, the backpack felt like it weighed about twenty-five pounds. I had two of these made, which would become unanticipated road-show companions for the next month.

The first time I displayed these mock devices was for the highest echelon of the FBI. Director Mueller's office had heard I had the devices and asked that I come up to give him and his sea of blue suits a briefing. I ended up sitting off in a small conference room for a couple hours to do just this.

Unbeknownst to me, while I was waiting for what I thought was a routine meeting, Mueller and the team were caught in our crisis center while the Watertown shootout was going down. Eventually, I got five minutes with him to go over the devices and explain their functioning. Mueller was a very hard man to read. He nodded a number of times, asked some good technical questions, and headed back out the door. I breathed a sigh of relief that I didn't do or say anything too stupid and figured that would be the end of things. Apparently, I had made more of an impression than I thought.

Less than a month after the bombings, then-UK prime minister David Cameron was scheduled to visit the FBI headquarters. This was the first visit of a UK prime minister to the FBI, and part of the express purpose of this visit was to learn about the US response to the Boston Marathon bombing. Mueller's team asked that I come up early with the mock devices so that I could brief Mueller again before he showed them to the prime minister. As my business card reads, "Have Bomb. Will Travel." I agreed.

I was greeted by Mueller's chief of staff in our crisis center. He rushed up and told me I would have five minutes. I told him that would be ample time to brief Mueller. He looked at me like a dog trying to figure out a quadratic equation. After asking me what I was talking about I told him I was informed my sole purpose was to bring a bomb and give Mueller a reminder of how it worked so he could take it to his briefing with the prime minister. He quickly informed me that I was mistaken.

Mueller was set to brief the case specifics on the Marathon bomb, but he wanted me to come in and personally brief the bomb to the prime minister and his staff.

Without a chance to collect my thoughts, I was rushed into the conference room. Mueller looked up and said something along the lines of, "Ah, the good doctor, have a seat and we will get to you in a few minutes." I don't recall much of the meeting as I was filled with some atypical level of adrenaline, but when called on to explain the bomb, I went through its functioning. Such briefings are made easier with props because the minute you pull out pipes with fireworks fuses sticking out and shiny pressure cookers all eyes leave you and focus on the cool-looking toys.

As I was leaving the briefing room, the chief of staff came up and noted that Mueller must really like me because he was very particular about who he lets brief with him. I was unsure if that was a good thing or a bad thing, but I soon found out it would be a recurring thing.

The deputy director (DD), who also sat in on my briefings, decided that I would make a good accompaniment for his various meetings with House and Senate staffers. At that time, the FBI was being asked some hard questions about why it did not catch on to the Tsarnaev brothers before they hatched their plot. There were indications they had traveled overseas and questions about whether or not they received terrorist training while on those trips. Hindsight being clear and focused, everyone was looking at political points to be scored. The DD was being called in front of committees filled with staff members to take some political heat. It seemed that I would be a good distraction during these events.

As a result of being a one-man traveling bomb show, I got to witness some nice heated exchanges between our DD and Congressional staffers trying to score political points. On cue, I would take out a backpack filled with bomb mock-ups and distract attention away from the political meanderings of the proceedings. I played one part educator, one part magician's waving hand, and all parts mystified observer.

The fun part was getting the bomb into the Senate building. I had to work with our Office of Congressional Affairs to get the bomb cleared. Eventually, Mueller became so enamored with the bomb he took custody of it himself. Having made two, I could hardly begrudge him a copy of his own. My final matinee showing of the device was with Mueller himself.

Having wowed a variety of House and Senate subcommittees and their staff, Mueller now wanted to go directly to the Senate budget committee. In short, it was time to do some chest pounding about the work the FBI did in the case and brief the senators, who control the purse strings. Mueller was bringing the bombs and he wanted me to do my one-man play after his briefing. So, dutifully, I showed up and waited in an anteroom outside one of the Senate chambers.

Eventually the briefing ended and Mueller sent for me. The two senators present were Richard Shelby and Barbara Mikulski. They both came over to a table where I had the device set out. I went through my spiel of how the bombs were constructed, how they were intended to be utilized, and how they functioned. This show always ended with me pressing the transmitter button and triggering the buzzer in the pressure cooker.

At the time of my explanation, Ms. Mikulski, a diminutive woman, was hovering intently over the bomb. Shelby, being larger, was a bit further away. By this time, I had done this show so often that I forgot one trivial aspect of demonstrations. Before I pressed the transmitter, I failed to warn the spectators that it would trigger a mock detonator (a.k.a. a loud buzzer in the pressure cooker). Mikulski gained about three inches in stature as she jumped up and yelled out in surprise when the buzzer went off.

Shelby found this funny and was smirking as I immediately apologized for my omission. I glanced over at Director Mueller's deadpan stone-carved face. "I'm really lucky he likes me," I thought. Otherwise, my career might have been over then and there.

CHAPTER 12

It Takes a Village

No one wants bombers to succeed. My role in the FBI for a decade involved trying to figure out how bombs were constructed and the manner in which all the pieces worked together. As time went on a more important mission took predominance: keeping bombers from ever assembling or delivering their devices. To do this everyone can play a part. It seems fitting to end with some perspective on how bombings have evolved and what an informed public can do to work with teams like mine to prevent their occurrence and to stay safe.

Bombers' motives are timeless. Some are as basic as those we see with everyday violent crime. Greed, envy, lust, and just about every other deadly sin (sans gluttony) can all play a hand in motivating bomb builders. For terrorist bombers, the motives can be more nuanced. Their attacks often result from what they view as oppression by a force much stronger than them. We know that historically, many bombers attack out of anger against a perceived power, be it governmental or industrial, that is somehow viewed as a threat to something they hold dear. The item under perceived threat could be their future, their children's future, their culture, or their religion.

Labor disputes in the US at the turn of the last century were the result of workers lashing out against the wealthy class trapping them in a life of poverty. This conflict eventually saw the wrath of bombers shift focus from the rich robber barons to the government officials who empowered them.

Ortie McManigal, described earlier, would be far from the last serial bomber our nation would face. As history moved forward serial bombers came and went. McManigal was replaced by George Metesky (the Mad Bomber of New York). Metesky was himself lost in the annals of time and would see Ted Kaczynski (the Unabomber) take his place in the collective memory. At the time of this writing, it may be Cesar Sayoc and his devices remembered by most. Without doubt more names will rise and fall.

All of these individuals shared some very common traits. Anger precedes acts of violence. Bombings are a logical extension of such feelings of rage. Oftentimes someone near the bomber sees the building of this emotional grenade. It is something to always be aware of as an indicator of potential concern. Typically, there is no need to be suspicious of a neighbor who goes on a rant about the IRS in April, but it might be worthwhile noting one who spouts escalating vitriol year-round and has started stockpiling chemicals and electronics in his garage.

The FBI is never going to be the first to notice a bomb plot start to develop. In all likelihood local police won't see it coming. Everyone's best hope still resides with folks who have a gut instinct that something is not right with the actions of a family member, acquaintance, or customer.

These days I spend a great deal of my time working on training retailers, hospitality-industry security specialists, industry chemists, academics, and countless other groups about warning signs to be cognizant of that might indicate a potential explosive threat exists. There is no way of telling where the next bomb might emerge. Uncontrolled anger is an indicator on a larger checklist for everyone to understand.

The bombers' choice of venues has not changed substantially over the years. Historically, a certain cross-section of bombers has focused their animus on locations where crowds gather. These individuals are driven by the horror of the body count. This evil subset of bombers directs their attacks against innocent civilians.

The first attack directed at a celebratory civilian gathering actually occurred over a hundred years ago in 1916. At the time the event also constituted the deadliest terrorist attack on US soil. With WWI escalating in Europe, sentiment in the US ran the gamut from isolationism to calls for America's entry into the conflict. A parade was planned in San Francisco to "boost patriotism and compel leaders in Washington, DC, to increase defense spending." The event, which drew approximately one hundred thousand spectators, was one of the largest held in the city's history.

During the parade, witnesses noticed an individual set down a suitcase on the sidewalk against the wall of a saloon along the parade route. One civic-minded bystander even warned the man to be careful lest his suitcase get stolen. The man vanished into the crowd, and at 2:06 PM a powerful explosive charge hidden in the case detonated. Ten people were killed and forty-four were hospitalized, in what became known as the Preparedness Day Bombing.

A six-year-old boy would have the flesh seared from his legs by the attack. This boy would become the first documented child casualty of a terrorist bombing in the US. Sadly, children would go on to become intended targets of such terrorist attacks as the Oklahoma City Bombing.

Eighty years after the Preparedness Day Bombing, Eric Rudolph also chose a gathering of celebrants for his infamous Centennial Olympic Park bomb attack (the largest of several). At the end of the first full week of the 1996 Summer Olympic Games in Atlanta, Georgia, Rudolph made his way into the Centennial Park venue carrying an ALICE (all-purpose lightweight individual carrying equipment, favored by the military) pack containing three massive pipe bombs filled with smokeless powder each encircled with masonry nails. Rudolph, like others determined to maximize their carnage, focused the effects of the blast by placing the pipe bombs atop a thick metal plate.

With a timer set to a fifty-five-minute delay, Rudolph set the bombs down beneath a park bench and departed the venue. He attempted to call in a warning to 911 after making his escape. He was hung up on by the first operator and was disturbed before he could finish his second call. The 911 call information would never make it to security officials in Centennial Park; however, a security officer (Richard Jewell) noticed the abandoned backpack and authorities began to clear the area. At 1:20 AM the device exploded. Two fatalities were attributed to the device.

Like so many before him, Rudolph was angry with the government, most notably over abortion laws, but he was clearly "anti" a lot of things, according to FBI records. His motive was to shut down the Olympics, which would be the ultimate affront to the country hosting the games. Atlanta had created Centennial Park as a central gathering location for visitors and spectators during the Summer Olympics. The park served as a high-profile target as it contained sponsor exhibits, and hosted entertainment and medal presentations. The Centennial Park bomb was the first in what he had planned as a multi-night campaign of bombs. Rudolph had four other bombs assembled and ready to plant. His displeasure with how the first attack went caused him to abandon his plot and detonate the remaining four devices to destroy them.

The theme in both of these attacks separated by eight decades is the abandoned package. Backpacks, briefcases, bags, boxes, all sitting in the open in a crowded venue should be treated with suspicion. Bombers never leave just a couple of sticks of dynamite, wires, and a clock attached to them. When targeting a crowd, bombs will be concealed in an item not meant to elicit attention or fear. In any public gathering area, an eye should be kept out for such anomalies.

In any bombing, there is always a chance for a secondary device. As previously noted, Eric Rudolph brought this lesson home in his abortion-clinic bombing. God forbid anyone ever experience a bombing. In such cases law enforcement always assumes there are more where the first one came from. So should you. Move well away from any area where there has been an explosion, a bomb threat, a suspicious package, or potential explosives. Curiosity often draws observers into a deadly zone when bombs and explosives are concerned.

Also beware of multiple bomb threats. The Provisional Irish Republican Army (PIRA) would purposefully call in bomb threats to areas to watch how police set up cordons and command posts. This provided insight into where to place a real bomb the next time around.

Anytime a material believed to be an explosive is encountered, nothing in the area should be touched. Many explosives made by amateur experimentalists can go off with little or no warning.

In 2018 a man transferred a significant quantity of TATP he produced into his SUV. While returning to his house, the material spontaneously exploded, scattering remnants of the vehicle over a hundred yards and onto the rooftops of neighboring houses. Responding bomb techs found twenty-two more pounds of this clearly unstable material. Only by luck did no one get killed.

I cannot recall the number of times I worked with bomb techs cleaning out an amateur explosive lab after something had gone horribly wrong. In many cases, friends, acquaintances, or even close family members were well aware of the explosive production going on. In just about all of these cases the experimentation was written off as harmless fun or toying with "fireworks." The danger these materials pose cannot be stressed enough. In many cases, had the family intervened a life would have been saved. At the very least a crippling injury would have been diverted.

Speaking of fireworks, although not of military-grade power, perfectly legal fireworks can create a horrific bomb. Anytime someone is doing anything but lighting the fuse on a legally procured firework, gruesome things can transpire. Taking apart fireworks to make bigger blasts is not harmless youthful hijinks. If only one parent reads this and stops Junior from tinkering with the stuff that goes boom, I will consider it a victory.

I remember a video in a case of a kid lighting a large "firework" he had assembled from combining many smaller articles. Something went wrong with his lighter and the firework went off in his hand, removing it. The screams in the video haunt me to this day. A pill bottle worth of material will completely take off a hand. There is no "small" when it comes to explosives.

Many common household chemicals can be used to prepare bombs. These days anyone who has watched the news has heard of peroxides and Ammonium Nitrate. Everyone who deals with any consumer chemicals should be educated on how they can be misused. The London bombers from the subway and bus attacks in 2007 made many pounds of peroxide explosives. They actually bought out a beautician supply store of hair dye, more than once, to make their charges. No one took notice.

On the flip side, there are success stories that should inspire us all. In 2011, a terrorist attack being planned by Khalid Ali-M Aldawsari was disrupted by a vigilant company shipping agent.

Aldawsari, a citizen of Saudi Arabia, was admitted into the United States in 2008 on a student visa and enrolled at South Plains College near Lubbock, Texas. Aldawsari began to procure precursor chemicals and labware required to produce the explosive Picric Acid. Having procured two strong acids, he finally attempted to purchase the third and final chemical, Phenol. Aldawsari attempted to have the Phenol order shipped to a freight company so it could be held for him there, but the freight company, questioning the legitimacy of the order, told Aldawsari that the Phenol had been returned to the supplier. After returning the chemical, the company contacted law enforcement to report this suspicious purchase. This report led to a disruption of the plot and the arrest of Aldawsari. Something bad was seen; someone said something, and that something was stopped.

By the very nature of bombing investigations much of the narrative in this book has wandered down some dark hallways. It would be unfair to close the final chapter without saying that I have been blessed with some really memorable and fun opportunities by virtue of my position with the FBI.

One of the bomb techs I worked with for years ended up assisting the MythBusters during their long-running television show. They would come to him for advice on how to blow things up, and on occasion he would come to me. My favorite exchange with him revolved around an episode where the team was trying to determine if an explosive charge could literally knock a person out of their socks without killing them. Truthfully, this is the kind of thing that I would never really have taken the time to cogitate about.

Posed with the question, it was fun to research blast pressure lethality and pressure distances. It was easy to find tables and figures outlining what pressures a person could survive from a bomb blast. Not so easy was finding literature on how fast a body has to move to leave socks behind.

Each time I came up with a survivable distance from a bomb blast a couple days would pass before another call would come in from the bomb tech telling me they wanted to use a bigger bomb. Having watched the show for years, this did not come as a surprise. Eventually, they settled on a fairly large charge of five hundred pounds and were able to launch a mannequin out of a pair of socks. The myth itself was busted: you could not survive an explosion that knocked your socks off.

As reward for my evenings of calculations I asked to be listed in the show's credits as a technical advisor. I could not get paid for the advice, but having official recognition at the end of the show was priceless. Having watched the show for years with my sons it was great to see the look on their young faces when my name scrolled past. That two seconds brought me more credibility than any case I could have ever worked in their eyes.

My knowledge of esoteric historical bombings, which has no doubt become painfully apparent by now, also secured me guest roles on the TV shows *America Unearthed* and *I Was There*. The *America Unearthed* episode recreated and studied the Haymarket bombing of 1886, whose trial transcripts I had found years earlier. *I Was There* focused on the Oklahoma City bombing which opened up this book. I had done TV interviews before that, but never saw the sausage-making that goes into a television production. As an added bonus to my TV star fame, I am now the proud owner of my own page on the IMDb website.

Apart from scattered TV appearances, scaring senators with bomb replicas, and warning everyone else into avoiding hardware, pool, and drug stores, what does a Senior Explosives Scientist do day to day? If nothing else I want to stress that I don't solve every bombing case that comes through the FBI Lab's door. When a truly complicated case emerges, I might work with bench examiners to conduct research and analyses that might be beyond those typically applied. I spend hours studying the most obscure aspects of explosives in the hopes of uncovering that one elusive insight that may help ensure a bomb tech goes home at the end of a difficult day. Most importantly I mentor those who will replace me.

The secret to success is to hire people smarter than you and do everything you can to make them shine. I have been blessed to be surrounded by a team of motivated and talented people. Some days I may wish to strangle them. Like when they put a handful of party poppers above my lab door which rained down around my feet in a loud crackling barrage as I was carrying in a tray of an explosive I just made. Other days I envy them and wish I could jump back into the fray.

Knowledge is not power. It is responsibility. Many people have trusted me with their lives over the years. I owe it to all of them to ensure those who follow me are smarter and better equipped to handle the challenges yet to come. There is a bittersweetness to feeling oneself become obsolete while striving to bolster those who will carry the flame forward. When the time is right, I will vanish into a high desert sunset secure in the knowledge I played a role in a timeless story that will unfold in chapters yet to be imagined.

Epilogue

It was a perfect spring day in 2020. The sky was clear, temps were in the low seventies. The day would have been glorious, with one exception: along with a good portion of my fellow Americans, I was under lockdown due to the uptick in COVID-19.

As I worked from home on various threat assessments and technical publications, an update from our local National Explosives Task Force (NETF) dropped into my inbox. These updates usually are fairly bland and contain stereotypical incidents of people being caught doing dumb things with routine bomb-making materials. I figured it would provide a brief respite from the paperwork in isolation.

But something stood out. For the first time in a long time, a bombing story gave me pause. Having thought I had seen everything, circumstances conspired to slap my jaded perceptions down with some authority. The story read:

> *On 24 March 2020, the FBI executed a sting operation against a 36-year-old male at the conclusion of a months-long domestic terrorism investigation. The suspect arrived at the scene to acquire an explosive device for a vehicle-borne IED (VBIED) he intended to detonate at a Belton hospital that was providing*

critical care during the ongoing COVID-19 pandemic. The device brought to the sting by agents was inert. When FBI agents attempted to make the arrest, the suspect, who was armed, was shot and later died at a local hospital. According to the FBI, the suspect was a potentially violent extremist motivated by racial, religious, and anti-government animus, but at this time, there is no publicly available information linking him to extremist groups. The suspect spent months actively planning to commit an act of domestic terrorism with an explosive device that would cause mass casualties, and he took steps to acquire materials to construct a VBIED. The suspect originally considered targeting a mosque, synagogue, or school with mostly African-American students. Due to the ongoing healthcare crisis from COVID-19 and his frustration with the Government's lockdown measures to stop the spread of the disease, the suspect chose to accelerate his plans and target a hospital instead of one of his original targets.

I sat and contemplated this for a while. Over decades of bombing investigations, I have noticed a slowdown of domestic incidents when our nation undergoes shared hard times. Yet, here in the midst of perhaps one of the most difficult times I can recall, a bomber not only wanted to start the typical race/class/religious war with an act of savagery, but he also set out to target those already afflicted by the ongoing tragedy. Perhaps one of the reasons I enjoy horror movies so much is that it is a relief to get lost in fanciful tales of fictionalized monsters. The real ones I all too often encounter are far more psychologically draining.

Within the same bombing incident update another story jumped out at me. It occurred a day after the horrific disrupted plot and stood as such a beautifully humorous counterpoint to the first incident that it was almost as if the universe planned it so. A simple tale that briefly parted some of the clouds obscuring the day's previous brightness noted:

Epilogue

On 25 March 2020, at approximately 1615 hours local time, the Eastpointe Police Department responded to reports of shots fired in a residential area. At the scene, police found evidence of a significant explosion that damaged two homes. No injuries were reported. According to police, a resident constructed and detonated an IED in an attempt to kill rats. Officers reported that no rats were killed, and the resident suspected of constructing the device was arrested and faces multiple charges.

My job can be heart-wrenching one moment and so blithely absurd the next.

My fellow bomb techs and investigators witness acts of carnage. We see evil unleashed in waves of pressure and shredded metal. But just as we start to convince ourselves of the world's evil, we are confronted with such idiotic, slapstick spectacles that it is almost tempting to look for the hidden camera filming our reactions to such a contrived theater of the absurd. And so is the way of the bomb.

Acknowledgments

You should write a book. I can't recall the number of times I heard and dismissed that old trope. For years I would politely nod and brush off the sentiment with a polite "Maybe someday I will." Then someday came.

Over the years I have been the subject of numerous interviews and articles, which might leave an audience thinking I had been personally behind solving every bombing case to cross the path of the FBI. Nothing is further from the truth. I have been graced with being part of a fabulous, dedicated team of bomb techs and scientists whose mission it is to conduct forensic analysis on explosive and bombing related cases. When I speak of "my team" it is not in reference to a team I lead like Captain America, it is in reference to team I work as part of, and in service to.

Contrary to overhyped descriptors used to capture my essence in the past, I am not elite (a word I despise), heroic, or a modern-day Sherlock Holmes single-handedly waging war against a relentless foe. I have worked with plenty of people smarter, braver, and almost as aesthetically pleasing as me. This book was inspired by a desire to tell the human story behind my job and to highlight the astounding work of the community I support.

When my sister was contacted by an old publishing friend, Heather Jackson, asking if I would be interested in writing a book, for reasons above, part of my reticence had already crumbled. With that my first acknowledgment goes to Heather (of the Heather Jackson Literary Agency), who knocked me off the fence and into the briar patch of the publishing world. I decided to write this tome not only to give voice to my reality, but also to embark on an adventure. Heather has been both travel guide and navigator through the rocky shoals of this odyssey.

I have spent the past couple decades writing exclusively for police, firemen, bomb techs, and the military. My co-author and sister, Selene, on the other hand, has spent a career distilling down a wide variety of technical subjects for the non-technical reader. She deserves the credit for taking my voluminous stream-of-consciousness diatribes and gore-filled ramblings and molding them into something that comes across more manuscript than manifesto.

Any author who claims sole writing credit is doing a great discredit to book editors. I was fortunate enough to have brilliant editors, copyeditors, and other support team members who made this manuscript better every step of the way. Accolades need to go to Katie Benoit, Madeline Sturgeon, Gretchen Young, and Clayton Ferrell for their careful eyes and sharp editorial assistance.

The people above helped create the book. It is family that creates the world worth saving. My father, Ronald Yeager, who dedicated his life to obstinately teaching others, whether they wanted to learn or not, instilled in me a love of storytelling, writing, and educating. My mother, Jane Yeager, always had a wonderful way of looking at the world through a prism of dark humor. It is this grim joyfulness that imbues the subtext throughout the book. Years ago, I recognized my parents at a huge award ceremony and noted, "Nature or Nurture: You're to blame either way."

Acknowledgments

I signed up for the FBI, but that act conscripted my entire family into a life of sacrifice they have graciously borne for more than twenty years. Plain and simple, none of this would have been possible without them. My wife, Deborah Yeager, has provided an anchor of sanity, stability, and love for this whole insane journey. My job often took me away without warning and occupied more time than healthy on occasion. Thanks to Deb, my two sons—Jared and Alec—have grown to be principled, responsible, and solid young men instead of the feral miscreants I know lurk beneath the surface. I am proud of both of them—and indebted to her.

To the men and women of the Explosives Unit both mentioned and referred to in the book I cannot express how proud I am to have numbered myself one of your ranks. To the FBI Lab staff working to shine a light of knowledge into dark deeds, the nation owes you more than they can ever know. To the FBI family protecting the constitution with quiet dignity and keeping the American people safe, my deepest thanks.

A word about the stories. I never intended to write a book. As such I never took notes as life flowed past. With age I have become fully aware of the foibles of memory. In every story I have done my best to recount the details as accurately as possible. I ask forbearance for those on scene who might remember a slightly different sequence of events.

My field passes on its stories as part of a rich, often beer-soaked, oral tradition. Fortune shined on me and provided a venue to impart some of mine to the world. It is with the utmost of respect to all the bomb techs and Explosive Ordnance Disposal technicians out there who have so many astounding stories of heroism, sacrifice, and adventure who will never get a chance to share their tales that I give my final acknowledgments. Serving in your ranks has been the honor of a lifetime.

About the Authors

Dr. Kirk Yeager received his BS in Chemistry from Lafayette College and PhD in Inorganic Chemistry from Cornell University. He worked as a research scientist and became the associate director of R&D at the Energetic Materials Research and Testing Center (EMRTC) in Socorro, New Mexico. While in the Land of Enchantment, he also held the position of adjunct professor in the New Mexico Tech chemistry department. For ten years he served, as a physical scientist and forensic examiner for the FBI Laboratory's Explosives Unit, where he deployed as a bombing crime scene investigator to dozens of countries. Currently, Dr. Yeager is the FBI's chief explosives scientist. Dr. Yeager has nearly thirty years of experience with improvised explosives and IEDs. Over the course of his colorful career, he has served as a subject-matter expert for the National Academies of Sciences, worked as a technical adviser for *MythBusters*, and been the subject of a feature article in *Popular Mechanics*. He is an avid geocacher and holds the rank of black belt in Danzan Ryu Jujitsu. His academic prowess is surpassed only by his charm and humility.

Selene Yeager is Kirk Yeager's sister. At least that's how high school teachers referred to her, right after "Ron Yeager's daughter," until everyone realized that she was a little bit like them...and a whole lot of something different. Terrible at chemistry and good at sneaking out of the house, she was single-minded in the pursuit of two things: riding a bike and expressing herself in the written word. The former has taken her around the world. The latter has helped her build a career in storytelling and journalism and as author, coauthor, and contributor to nearly thirty books, granted her a nomination for a National Magazine Award for excellence in service journalism, and given her the great honor of bringing her brother's work to the world through this book.